The SolarCycle Diaries
By Susie Wheeldon

Dedication

This book is dedicated to the many incredible people around the world who fed us, sheltered us and laughed with us along the way.

You made our journey un-missable and unforgettable.

The SolarCycle Diaries

"So, um, tell me again why we're cycling through the Sahara in June..."

In 2010 Susie Wheeldon and Jamie Vining rolled back into London complete with stings, bites, leathery skin and thighs of steel. With their co-rider, Iain Henderson, they had set out nine months earlier on a 12,000 mile round the world bike ride, promoting solar power and being tracked using solar nanotechnology.

Only, as Google Maps and your thumb are not the most precise tools with which to plan a global circumnavigation, they had instead ridden 13,500 miles across North Africa, past the Himalayas and through some of the Earth's more pungent road kill. Braving blizzards, tackling thunderstorms and getting blown off the road, they had been chased by dogs, attacked by killer bees and rerouted by the bubonic plague.

Their quest to promote all things solar had seen them followed by the police, accosted by goat herders and giving a talk—from a dog kennel—somewhere in the Deep South of America.

From the painful to the bizarre, this in an account of their journey through France, Tunisia, Libya, Egypt, Jordan, Syria, Lebanon, Turkey, Iran, Turkmenistan, Uzbekistan, Kyrgyzstan, China and the USA.

It was one hell of a ride.

Susie Wheeldon
With Jamie Vining & Iain Henderson

Acknowledgements

Thank you to family and friends for the support you gave us every step of the way, as well as all those who profiled the trip as we pedalled around. This acknowledgements page is book related but I cannot express how much it was appreciated. Without you, none of this would have happened.

In the making of The SolarCycle Diaries, though, I would particularly like to thank Nicola Leslie, Katrina Vines, Charles Lamb and my parents for their services to editing and content. It was a big ask and I promise that in return I will learn how to use apostrophes.

Izzy Way and Ollie Stewart, thank you for the incredible front cover, Ollie especially for the original design and for being more patient than should be humanly possible. I will be buying you a copy of Ground Hog day for Christmas.

Blake Roseveare and the Lambourne's, thank you for taking in waifs and strays and providing me with a place to write.

Sam Mellish, photography school was much appreciated as was Meena Rajput's IT training; more accurately described as, fixing all of the formatting and images since it was less painful than watching me try. I am still suspicious that you've 'accidentally' deleted me from the launch party though... if you are reading 'The SolarCycle Diaries by Meena Rajput', call security.

On the topic of launch party and press, thanks also to Hannah Bailey, Timmy Manners, Ruth Carruthurs, Lara Jelowicki and to Quiksilver.

Anna Wells and SolarAid, from the very beginning you have been absolutely incredible and I am eternally grateful for all of the ideas and support. Tea is always brewing in Tropical Peckham.

Pete, my brother, thank you for the 'Note about the author' and for general encouragement, it meant more from you than it did from anyone else. (Though I am still pleased you took out the bit that made me sound like a whale.)

I would like to make a special mention to my co-rider, Iain Henderson, for his diary inserts and for providing many of the stories held within these pages—if only every day could have you getting us into a hotel by pretending to be a Speedo model.

But most of all I would like to thank Jamie Vining. Not only because he was the one who actually got me round the world but also because he took nearly all the pictures in here and drew all the maps at double speed

"Two days? Ah well, whatever I draw will be better than your attempt with a marker pen and a print out from Google!"

You are amazing.

Note about the Author

Susie is one of 'those' people.

You know, the people that 'normal' folk like you and I enquire after with a strange sort of indulgent smile on our face. We share that smile because we know the answer will be so unusual that we need to insulate ourselves from the strangeness of it by first reinforcing how we are both in the 'normal' club.

Normal to us is carefully building our little bubbles of friends and doing our sensible jobs and creating our normal families; routinely only going to the places we know are nice—and not too rough—with only the odd flash of 'daring' to brighten things up for a moment.

Susie doesn't really care about those things.

Because Susie isn't normal.

Susie's normal is the stuff you or I would never ever think was good idea, even just before closing time. Susie's normal is walking from Peckham to Camden for fun, in the rain. Susie organises pop up pantomimes and Top Gun volleyball parties. Susie runs over buildings and in the Marathon de Sables, then builds websites that may, or may not, save the planet. Susie gets a job as a rigger at Glastonbury then successfully gate-crashes Bono's VIP after party.

Susie's enthusiasm is utterly infectious. Susie talks to random people every day in London... and they talk back.

In 2009/10 Susie decided to cycle around the world.

That is her normal.

The bare facts are these:

The trip

13,500 miles, 14 countries, 4 Continents, 9 months.

The transport

Bicycles (and 2 plane journeys under significant duress).

The trip leader

Susie Wheeldon: Aged 30, 5 feet and ½ inch tall, 7 stone. No genetic predisposition towards endurance. Likes flamingos.

Team

One Jamie, half an Iain, a bit of a Charles, and others as required by law.

Languages spoken

English (sort of), wild gesticulation and 'helpless blonde'.

The rationale

To cycle through the deserts promoting solar power as a solution to climate change (a faintly disguised excuse to get a tan).

Previous overseas cycling experience

Not nearly enough.

I'm not like Susie.
I'm her brother.
And for the first time ever, I've realised I'm a bit jealous.

Peter Wheeldon, 10 November, 2011

England

Estimated Distance: 121km
Actual: 127km
Sunlight Hours per Year: 1000—2000

Riding Tunes:

"You've got to change the world
And use this chance to be heard
Your time is now."

Butterflies and Hurricanes, Muse

Susie's Diary—May 2009 (EU Solar Day)

It's 6am, London, my friend's living room.

I am panicking. The website is not up. I haven't packed.

I haven't—most worryingly—got a sports bra.

I am a 30 year old woman. 9 months leaning over with inadequate support is not going to do me any favours.

I call Sophie. "Soph, you know you said you would help with any last minute problems..."

***9.15am** I am outside Jamie's house. I am late. We are late. I call up. Be down in 20 minutes, do you want a cuppa? 20 minutes? 20 minutes? Why would I want tea? I am too stressed for tea. Why isn't he ready when we're already late? I haven't put any make-up on and there will be photographers. I fume silently. And then unpack my make-up bag. I am putting on mascara in the road whilst simultaneously brushing my hair with my fingers. Buses ricochet past, missing me by inches. Jamie eventually emerges and we pedal off. The toothpaste and eyelash curler I've left on top of my pannier fly off at the Wandsworth intersection. Keep going. There's no time. I can probably live without an eyelash curler in Uzbekistan.*

***10am** We get to City Hall picking up Iain on the way down Tooley St. Iain is the third member of our rag tag team. He too is late after last minute problems renting his house out. We grab bagels and coffee while branding our bikes with the stickers from Nokia and our solar sponsors. Organisation is not at a premium.*

***10.30am** Various friends, family and photographers arrive and we are hustled and bustled through a series of shots and muddled interviews. Totally out of our depth in every way. A bird eats the majority of my breakfast. Bloody bird.*

11am *The Mayor of London, Boris Johnson, explodes onto the scene for a hurried 10 minutes. He is as knowingly blustering as we'd hoped. His cries are immediately stifled by his harried press secretary.*

"These solar panels are amazing. We should get some for the building."

"Erm—we have them Sir. That's why they're here..."

11.15am *We leave. Take One. i.e. we aren't actually ready to leave but the photographers want to take pictures. We do a flyby of City Hall. The security guards are not happy. We are leaving for our round the world bike tour but aren't technically supposed to be cycling near the building.*

11.30am *Timmy and Nial, two friends joining us for the start of the journey, finally turn up. My saviour, Sophie, throws me a bag from Marks and Spencer. They only had an A cup. It'll have to be a push up!*

11.32am *We leave for the second time after a quick round of speed hugs. We are devastated to have to go when so many people have come along to say goodbye.*

11.33am *I fall off my bike as we set off.*

11.37am *We stop round the corner to finish packing.*

12.15pm *I cycle into Iain (for the first time) and graze my knee.*

12.16pm *We are still 80 miles from Dover, our destination for the night.*

Summer 2007—almost two years earlier

I can't remember the exact point at which the plan to cycle round the world was conceived, but it was sometime over the long hot summer of 2007. A few weeks earlier I had shuffled onto the soggy tarmac of Heathrow after my first big adventure challenge: the Marathon des Sables. A 150 mile, 50 degree, fully loaded six day run over the Merguza dunes of the Moroccan Sahara. I was bored—flat after the whirlwind of excitement and adrenaline. No longer deliberating how close to the race campsite it was decent to pee, or playing ever more indecent versions of 'Would you rather...?'

My despondency was exacerbated by a reluctant lethargy. With bandaged feet ripped apart by the incessant grinding of sand—my flapping, blistered flesh amalgamated by the surgeon into three painful wounds—I was still unable to walk. After years of running, leaping and climbing I was now relegated to wearing slippers and wincing around my kitchen. I tried to busy myself and ignore the lively chatter floating in from the street outside.

Idly, I sat down at the computer to pass some time. With two friends I ran a sustainable living website. It was a tongue-in-cheek jumble of information, profiling solutions to climate change and highlighting various actions and technology. I had taken a solar charger to the desert with me and its simple, effective application had made me curious about harnessing the sun's power. I made a strong cup of coffee, curled up on the kitchen chair and absent-mindedly set about some research.

The Desertec Project

Concentrating Solar Power (CSP) is a process whereby mirrors concentrate direct sunlight onto one area to create heat. This heat is then used to raise steam to drive turbines and generators to make electricity.

It has been calculated that, using current day technology, less than 1% of the earth's deserts, if covered with CSP plants, would produce enough electricity for the whole world. Several of these plants are already in use around the globe.

Energy from them can be transmitted thousands of miles by high voltage direct current cables. Updating electricity networks and paying for CSP plants needs international cooperation and investment.

The Desertec Foundation is working to promote and facilitate such collaboration.

15 May 2009—Dartford High Street, England
Day 1—22km

Nothing beats a Ginster's pasty on the pavement outside Asda, drizzle wringing through the dishcloth of the clouds above. I perched my vacuum-packed pastry on top of an overflowing dustbin and grinned at the guys.

"Come on. It's not *that* far."

We had reached Dartford, and already the decision to plot the route using my thumb and Google Maps was coming under fire.

"When you say 80 miles, is that, in fact, code for 90 miles?"

"Well, it's still technically in the 80's... Anyway, don't blame me. Iain was meant to be in charge of route planning. Only on his map he missed off various roads and all of America."

I was determined to get to Dover by nightfall. Sadly, after a series of last minute panics, a few wrong turns and Nial getting a puncture, things were not exactly going to plan. It was to be several more protracted hours before, blinded by the driving rain, we slid uncontrollably down a vertical cliff face to the welcoming lights of our dockside bed and breakfast. Nial's spate of bad luck had continued. His panniers had enthusiastically detached themselves from his bicycle on a number of occasions and his frustration had risen exponentially with every hill we encountered before he finally broke and thumbed a lift in a passing transit van. Unfortunately for him, this brief respite was not the blessing he might have hoped.

"Hey guys, that was awesome. Hope you're still loving the cycling. I'm about 10 miles down the road. Pick me up when you get to Canterbury."

"Erm, Nial, don't get upset but... we're not going to Canterbury."

Jamie and I defected on a rescue mission leaving Iain and Timmy to continue at a more sedate pace. Timmy, as an old snowboarding injury in his knee had begun to twinge, and Iain, as he was surreptitiously hiding the fact that he had never cycled so far before. It was well after night had fallen, and after the gusts turned gale force, that we turned off the main drag.

Without the onslaught of lorries to protect us, a staggeringly vicious wind side-swiped us along the cliff tops and hurled us down the perilous descent. Skidding to a halt at the bottom I peered up at the un-amused faces of my sodden parents standing sentry in the darkness. Having found Iain and Timmy moments earlier, my mother had immediately taken charge and busied the bedraggled pair indoors. Now we too were rounded up with military precision and marched inside.

"This really is not on! We want to make sure you leave the country dear. You're drenched. Lock your bike up. Give me your clothes. Have a shower. I'll send your father out for pizza."

KIT LIST

Bike[1]
Surly long haul trucker frame
26 inch wheels (easiest to repair worldwide)
27 gears
Schwalbe Marathon Plus tyres
Shimano double sided pedals
Shimano r550 brakes
Brooks saddle
2 x rolls handle bar tape

Camping equipment
Terra Nova Lightweight tent
Cumulus Sleeping Bag
Primus Stove
Set of 3 pans
Mug
Spork
Camping Towel

Clothes
4 x t-shirts
7 x knickers
3 x socks
1 x jumper
1 x trousers
1 x thermal
1 x raincoat
1 x shorts
1 x sundress
1 x sarong
3 x bras
1 x sunglasses
1 x cycling gloves
1 x New Wave cycling shoes

---

[1] My bike was put together by James Thomas at Bicycle Richmond. I asked for a bike to take me round the world. It did. www.bicyclerichmond.co.uk

Books
The Solar Century - Jeremy Leggett
Ten Lessons from the Road - Al Humphreys

Maps
Tunisia, Libya, Egypt, Jordan, Syria & the Lebanon, Iran, Central Asia, West / Central China

Toiletries
Toothbrush
Toothpaste
2 in 1 shampoo & conditioner
Soap
Deodorant
Moisturiser
P20 sun lotion
Hairbrush
Hair ties
Hair clips
Tweezers
Cover-up
Face-powder
Mascara
Razor
Tampons
Ear-buds
Nail clippers
Nail file
Insect repellent

Medical
Medical kit
Malaria tablets
Plasters
Aspirin

Food and drink
4 x emergency energy bars
Water purification tablets
Iodine Solution

Spares and tools
2 x spare inner tubes
1 x multi-tool
Patches
Glue
Spare chain links
Spare nuts, bolts, screws
Oil
Tyre Levers
Pump

Miscellaneous
Small bag of washing powder
Pen
Notebook
Scissors
Masking Tape
String
Wallet
Passport
Torch with $100 emergency money hidden inside
Plastic container
Solar shower

Communications/Navigation
Nokia N90 phone
Blue tooth keyboard
Chargers, leads and adaptors

Bags
Small bag below seat
2 x Small Altura panniers

Silicon Solar Panel
1 x silicon solar panel with custom built metal bracket - affixed
to handle bars

Flexible Solar Panels

Our panniers were custom made with inbuilt nano-technology solar panels and used to charge our phones. This was designed to give us communications and GPS anywhere in the world.

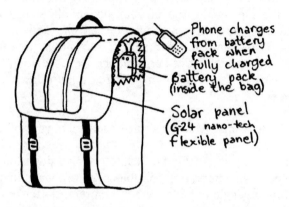

Phone charges from battery pack when fully charged

Battery pack (inside the bag)

Solar panel (G-24 nano-tech flexible panel)

Kit weight:
Susie: 35kg
Iain: 45kg
Jamie: 45kg

16 May 2009—Dover, England

Day 2—126.86km

"TIMBER!"

Designed to allow for upwards pull, clip-in cycling shoes attach to a metal clasp on the top of your pedals allowing you to maximize momentum throughout your stroke and increase efficiency. This graceful symbiotic movement minimises muscle strain and *improves performance by up to 25%, using the full range of leg muscles in both the quadriceps and the hamstrings.*[2]

Or—in simple terms—your shoes are attached to your bike.

If you happen to forget this fact while cruising down Balham High Street in rush hour, it is possible, even probable, that you will fall into a gaggle of harrowed commuters, clatter to earth and lie there, stranded like a fitting guppy, until a kindly soul removes your shoe from your foot and releases you from your prostrate anguish. It is then equally as likely that you will subsequently fail to remove your other foot from its metal prison and be forced to push your steed barefoot, shoes dangling accusingly, to the nearest pavement.

My morning's customary low-speed tumble was, naturally, on the heavily congested entrance ramp as we boarded the ferry to France. Pogoing in the horizontal-fully-attached-to-the-bike crash position, I scrambled awkwardly away from the backed up lorry drivers while the guys laughed too heartily to help.

Thank God my parents hadn't seen.

January 2008—a year and a half earlier

"I expressly forbid you to go!"

---

[2] Livestrong.com

My father's words left little doubt about his feelings.

"You can't! I'm 29."

My mother looked on with relief. She had always promised to support my ambitions, no matter how outlandish, and was overjoyed that my more sedate father had begun an unusually forcible protest.

"Well I want to find someone else to come. But if I can't, I don't see why I couldn't go on my own. I've spoken to two guys who've been on long distance cycle trips and they hardly had any problems."

"They were men."

"You can't object just because I'm female! Mum, help me out here."

"I can't speak for your father dear…"

April 2008—a year and a quarter earlier

I dragged my head off my pillow and scanned the room.

It wasn't mine.

Quick assessment told me that I was still wearing—at least some of—my clothing from the night before. Events came slowly flooding back. It had been an incredible wedding. Lots of food, lots of drink, lots of dancing, lots more drink. The weary thud of an oncoming headache made itself felt behind my squinting eyes while the sunlight pierced my brain. I crawled out of bed in search of water. I'm sure Jamie agreed to come on the trip with me yesterday. He did. He definitely did.

Was he joking?

Gathering the troops

Jamie, an old university friend, lived in France. Months earlier he had agreed that if I was planning a round the world cycle he would join me for the section through Europe. Four weeks after the standoff with my father we had ended up at the same wedding, where he flippantly mentioned there was no reason not to do the whole trip. He was self-employed, building houses in Normandy; the houses weren't going anywhere.

Two seconds later, the tequila came out.

I was over the moon. Though I had remained steadfastly nonchalant, quietly I was sh*tting myself at the thought of a solo expedition; my mother's suggestion that she could follow me in the car was an even more terrifying alternative. Jamie was not only one of the most laid-back people I knew but—as an added bonus—would allay my parents' fears of brutal molestation just outside the M25.

The next month I bumped into Iain, a family friend. He was deliberating a career change and considering a cycling venture to Australia.

We had a team.

The Team

Bolstered by a couple of pace setters, five of us set sail to France that morning.

Jamie: chief navigator; fluent in French; logistics expert; strongest, fittest, best at bicycle mechanics; more useful than the rest of us combined.

Iain: most able to find a nice hotel or decent restaurant; excellent—if inventive—storyteller: "Reminds me of the time I was scaling the Eiger, my friend Ed had broken his leg in a fall so I had him across my shoulders when, astoundingly, we chanced upon two bikini clad members of the Swedish lacrosse team..."

Timmy—joining us to Cairo: comedian; in awe of Iain's storytelling prowess; specifically the insight that any mediocre sketch can be turned into an immediate classic with the final punch line: "I mean, it wouldn't have been so much of a problem but I was tied to another man at the time."

Nial—joining us as far as his holiday permitted: for some unknown reason speaking in an Australian accent and, having bounced back from infuriation on day one, declaring everything to be "sensational".

Me: notionally in charge; in reality getting merciless grief for the constant inaccuracies in distance and falling off every 25 minutes; about to kill Timmy if he faffed about one more time.

France

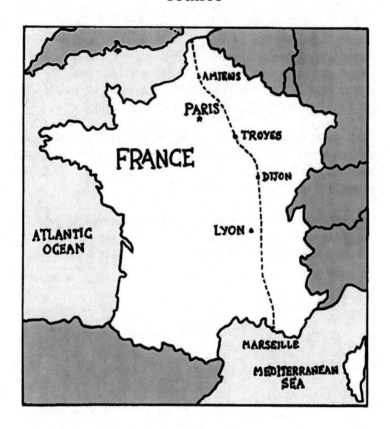

Estimated Distance: 1104km
Actual Distance: 1193km
Sunlight Hours per Year: 1000—2000

Riding Tunes:

"After chasing sunsets
One of life's simple joys
Is playing with the boys"

Playing with the Boys, Kenny Loggins

Calais, France—a stomach churning ferry ride later

You feel somewhat inadequate standing with your bicycle in the car lane for a ferry; lining up in formation with the other vehicles yet totally exposed to the elements. On the way off, though, it is another matter. On the way off, you get to indulge in every cyclist's favourite pastime: overtaking cars.

Sadly, the excitement did not last long. Burning past the traffic we followed the 'Toutes Directions' signs and headed straight for the emptiness of the sallow French countryside. Though the sky was ominous, the truculent weather had called a begrudging truce, and the gentle terrain did nothing to test our mettle. With cycling a national pastime, motorists gave us a wide berth and our incident levels dropped from a Code Red to almost negligible. We stopped intermittently to eat Muesli bars in grey lustreless towns and entertained ourselves by stretching inappropriately in front of passing caravans. Sometimes we spoke to a lone shopkeeper; sometimes there wasn't a soul in sight.

Host to the greatest cycle race on earth, neither France, nor we, were quite yet up to the Tour's great standards.

17 May 2009—Arras, France

Day 3—252.31km

"How close is Timmy to being ready?" I asked Iain with a weary sigh.

"Well, how do you define close?"

"As in, likely to be finished in the next half an hour?"

The muffled noise of crashing, banging and a profusion of expletives escaped through the doorway.

"Maybe you should have another cup of tea."

I walked over to Jamie and Nial hanging out by the bikes, shrugging off the first drops of rain.

"You guys alright this morning?"

"Sensational mate!"

"How many miles are we on for today, Mother Goose?"

"Oh you know Jamie, the usual 70..."

"So that'll be all the way to Marseille by lunchtime then."

"Has no one taken the map off her yet?"

20 May—Chateau Thierry, France
Day 6—483.31km

After the inevitable teething problems and the dull ride off the ferry, things soon began to gambol into a jovial adventure. We had spent a day in the gothic town of St Quentin and began to enjoy the more genteel aspects of a cycle through France. Mostly, wine. Albeit cantankerous, the weather had also improved since the aquaplane to Dover and Iain's unlikely tales were buoying us merrily along. Grand renaissance houses now peeped from hillsides and our bellies were filled with almond croissants and the odd injection of throat-searing Pernod. In a desperate bid not to jack it all in for a year in the champagne region, we made swift progress south and fell into a steady rhythm:

Clock up some miles before a mid-morning break (give each other grief), do another 30km to lunch (give each other grief), edge past the 100km mark (give each other grief, attempt to wee inconspicuously, get cycling shoe stuck in field), push on to destination (declare everything 'sensational', eat food).

Riding a bike is beautifully simple. No distraction, just deep breaths of fresh air and the feeling of muscles tightening and stretching. Nothing more complicated than finding food, finding water and finding somewhere to lay your head. On the open road I would drift in and out of daydreams, finally mastering the graceful push and pull of the pedals; while just behind me, head bobbing, shoulders jerking, Timmy was frantically swapping effort for strength in a bid to protect his injured knee.

No amount of frozen peas could detract from the fact that it wasn't getting any better.

22 May 2009—Dijon, France

Day 8—730.43km

The ride to Dijon was another of dreary clouds and underestimated distances. With despondency not only from Timmy, whose swollen knee was acutely painful, but the rest of us from general damp and exhaustion, it was a broken crew that finally rolled into the municipal campsite. Fortunately, Lady Luck had been waiting for a chance to arrive and rolled up just as lightning electrified the sky. As the fattest globules of water exploded between the hedgerows, we threw our bikes down, grabbed our panniers and heaved everything under the canopy protruding from the van of one Mr Mick Vining.

Mick, Jamie's Dad, had decided to visit us before we left France and settled on Dijon as his destination. For us it was a resounding choice. Diving under the awning, we took in deep lungfuls of tomatoey goodness emanating from two bubbling cauldrons and watched the torrent of water cascade from its roof to the grass outside. Mick whipped open a box of Merlot, threw on some pasta and 10 chorizo-warmed minutes later, life was unutterably better.

As an added bonus, we had made it to our first solar stop.

Solar Euromed

"Solar Euromed develops solar thermal power plants based on optical and heating properties of sunlight to produce electricity with the ability to include thermal energy storage to allow continuous production."—Solar Euromed

The following morning we were met by its CEO, Mark Benemarrze, as well as the Mayor of Genlis, the town in which Solar Euromed is based. Before leaving the UK I had been in contact with a number of solar companies, keen to see their work in action and to write about them as we blogged along the route. There had been a flurry of email exchanges with a number of organisations in which I agreed to do pretty much everything they suggested in a slightly over-excitable way. One of these was Solar Euromed.

"So, what exactly are we doing Suse?"

"Er, hmmmm, I'm not sure. But they sounded nice..."

Mission

Inspired by the pounding heat, mile after mile, day after day, on the Marathon des Sables, I had Googled, read and sunk numerous cups of coffee researching the power of the desert sun. I doubt it will come as a surprise to many that basic investigation confirmed its level as: High. Tents full of limp bodies attached to saline drips, a dizzying battle with dehydration, and the tragic death of a co-competitor on the 'Toughest Footrace on Earth', had hammered that home already. What I had been amazed to find, though, was the very real potential to harness this power and to use it as a renewable energy source.

Though there were a plethora of articles and documents on the subject, none were as comprehensive as those explaining the Desertec Concept. The Concept is simple; to harness the desert sun through vast Concentrating Solar Power (CSP) stations which then store, or directly transmit, electricity to towns and cities using high voltage direct current transmission cables. With energy losses of just 3% per 1000km, it estimates that the earth's deserts can easily service the 90% of our population that live 3000km from their borders.

The Desertec Foundation—organisation behind the Concept—was started by a group of European scientists called the 'Club of Rome'. Already impressed by CSP's potential, my desire to find out more was naturally exacerbated by the image of secret handshakes, Q-style gadgets and a bevy of bearded intellectuals. It was with a smattering of disappointment, then, that I found one of its founding members would not only meet me the following week, but would do so without even an inkling of a double-bottomed brief case.

Polly Higgins, put the kettle on, addressed my ignorance of high voltage cabling and excitedly helped revise our route from the Ukraine and Kazakhstan, past solar stations in Egypt and the Alps. Completely forgetting to consult the guys in my eagerness, I mentioned the direction change to Jamie in passing.

He paused more abruptly than anticipated.

"Let me get this straight. Instead of 'round the world' taking us through Poland, with women in short skirts, stunning architecture and cheap beer, we're now going across the Sahara desert and countries where the closest you get to a lady is a flash of ankle?!"

"Well—err—now you put it that way... but they don't have a lot of solar power in Eastern Europe."

"Then surely we should tell them about it!"

Genlis, France—back on the road

A ripple of flash bulbs greeted us as we rode up to the Hotel de Ville. We smiled broadly, pouted badly and pushed Jamie forward to answer questions for the local press. With absolutely no idea what was in store, we were duly bustled to the headquarters of Solar Euromed, guided through complex project demonstrations and stuffed with an array of petit fours.

Two hours later, slightly dazed and now back in front of the gathered throng, we bowed our heads as the town's Mayor proudly presented me, then Jamie, then Iain, with medals and a Gallic handshake.

What an incredible reception.

And a great honour.

Though, only 728km into our trip and battling nothing more than the vagaries of the French restaurant trade[3], we couldn't help but feel a little fraudulent.

The Route

Since its inception two years earlier, the trip had gone through a number of incarnations. After research into solar rickshaws and attempts to invent a solar powered bike, I had settled on the idea of using solar communications to record and track our self-propelled journey. With plans to promote the Desertec Concept, I also had ambitions to highlight other forms of solar power, lend our voices to the 'We Support Solar' campaign, and raise money for the charity, SolarAid.

Though the guys were fully behind these ideals, I was by far the most driven; constantly researching, talking all things renewable and emailing anyone and everyone who might want

---

[3] "Seriously, you close for lunch?" "But of course. We ave to eat ze lunch."

to get involved. In reality, both Jamie and Iain were on the expedition for other reasons.

Iain's motivations were the people and places we would find along the way; the romantic image of pedalling the Silk Route and the incredible Roman, Egyptian and Medieval ruins we would now be able to pass. He had an innate sense of adventure and a curiosity about the world. Fond of a good tale over a fine wine, he was happiest whilst undertaking any number of entertaining activities, but most especially those that would bolster his own, embellished, supply of facts and anecdotes. His eclectic career had already seen him paddle down the Thames, rampage through the Himalayas and eyeball a bull whilst attached to a friend by a six foot bungee cord. His addictive humour, encyclopaedic wealth of information and complete lack of shame made him an incredible travelling companion. Both easy and outgoing, he was the first person to make light of a situation or turn a tedious ten minutes into a side-splitting escapade. Dubbed 'Captain Flash Heart' within moments, the guys were in awe of his almost inhuman ability to sniff out a curvaceous woman, locate a fine restaurant or induce an unimaginably hilarious series of events.

Jamie was equally as compelled by the lure of adventure and the many bizarre stories we'd have along the way. He was, though, more driven by the physical feat of our undertaking than the sights and societies we would pass. He had cycled a lot as a teenager and was fascinated to see just how far he could push his body and how easily he would cope with the difficulties the trip would bring. Most importantly, Jamie was determined to find out whether or not cycling hundreds of miles a week would result in the mythical six-pack he claimed to have sported in his early twenties. He was dogged, determined and highly competent; something not so apparent in the rest of us. With an inherent ability to fix anything, find anything and channel MacGyver, he also came merrily equipped with a dirty laugh and the mischievous humour of a twelve year old. True, he might be navigating a complex intersection through the industrial suburbs of a busy French town, but he would simultaneously be contemplating the least appropriate place to ping open my bra strap—probably whilst talking to a policeman.

Though we had Timmy and Nial to ease us into the first leg of the journey, I could not have been luckier to have Iain and Jamie with me on my global cycling escapade. Not least as they had taken the new itinerary in their stride. No longer would we pedal the green rolling hills of Eastern Europe, past their perfectly proportioned female inhabitants, quaintly rustic bed and breakfasts and plethora of seasonal fare. No, we would instead head to the most desolate expanses of the world's least populated corners, ride across their barren terrain and test our nerve against the elements.

Our route would take us from the safety of France and Western Europe across the Mediterranean to North Africa, the Middle East, Iran, Central Asia, the Western Chinese desert and on to the U.S.A.

But, while we would no longer have the relaxing ride we might have otherwise found, at least I could assure them of one thing:

Challenge, adventure and a few impressive stories could not fail to follow.

23 May—Macon, France
                                          Day 9—883.58km

Back in Dijon, the first blow to the expedition was losing Timmy.

Having planned to join us until the Pyramids of Cairo, his knee just couldn't take him any further. With no time for recovery and the temptation of Mick's van metres away, reason finally saw through hope and he conceded defeat. It was with real sadness that we watched him roll out of the campsite, though probably more for us than it was for him. An impish grin spread across his face as he waved from the passenger seat and left us under a volley of abuse. While we would struggle on, he would retreat to a friend's nearby, lie in a hammock and drink Pastis.

Lucky bug*er!

Fortunately for the rest of us, the meteoric downpour had blasted away the clouds and we were finally able to relax into a steady pace under balmy sunlight and bluebird skies. Our route around Macon now took us down one of the 'Voies Vertes'— former railroads, haulage paths and waterways, which cross the nation. We gladly left the main roads that slice through the countryside and made our way down a converted canal, gliding past vineyards and chateaus, breathing the clean, crisp air like an instant tonic. Beginning to realise that a trip of this nature changed almost as rapidly as the weather, we settled happily into the peaceful holiday feel and stopped in a clearing by a babbling brook. Buying juicy cherries from a local bakery, we gorged ourselves, crimson toothed, as gentle apple blossom drifted softly across us.

"Jamie. Were you taking a picture of me having a wee again?"

"Only a bit. I was also trying to get the lady sunbathing with her top off…"

May 2009—Chanas, France

Day 10—1016.67km

In Chanas, we lost Nial too.

Although he had only ever been able to join us for a few days, it was still disheartening when he finally pedalled off to take the train back home. Not least because the morning he left we were accidentally located in a car park on the edge of town eating something masquerading as cereal, while the sun-burnt Englishman opposite gave a detailed description of his unpopular ex-wife. Nial's quick one-liners had elevated the dispirited atmosphere from morose to hilarious as he cheerfully packed the 'light reading' material Iain had bought him from the petrol station.

He would be sorely missed.

A fleeting melancholy permeated the depleted team. In such close company the mood of each of us had an immediate effect on the others. With inappropriate limbering, tag-team heckling and my near decapitation whilst swinging from a 10ft road sign; so far, the journey had been more mini-adventure than hard core expedition. With Timmy and Nial gone, I was nervous to see how Jamie and Iain would fare without the buffer of extra bodies.

Though I knew them both —Jamie from university and Iain from school days—they had only met each other twice before the start line.

Under the strain of the journey, would they get on?

27 May 2009—Salon de Provence, France
Day 13—1252km

As we headed south the mercury soared, bringing with it a taste of things to come. Iain's water intake—and outtake—took on epic proportions, and Jamie spent every possible second searching out the tiniest vestige of shade. It was long, hot, sweaty cycling as we pounded out the miles under the 35° temperatures. The extra exertion was, though, tempered by the extraordinary riding conditions. The Rhone had gouged out a pedal friendly gorge through the hills of the Massif Central, and the French roads were well paved with only a steady stream of traffic.

This was France as you dream it might be. Petanque in town squares, baskets hanging from ornate balconies and the fresh smell of croissants escaping through the doorways of early morning bakeries. We passed a day leisurely at a campsite in the ancient town of Montelimar before doing our best to cling to our bicycles as we flew towards the sea.

The 'Mistral' winds had arrived.

"Wooo hooo!"

The 60km ride to the Roman settlement at Orange, pencilled in for half a day, saw us screaming through the miles and eating our breakfast by the crumbling amphitheatre. By mid-afternoon we had battled our wilful bikes, hurtled ourselves round corners and prayed for survival the rest of the way to Salon de Provence. Blown through the doorway of the local tourist office, faces like sand blasted Botox, we caught our breath and released Iain on the situation.

An operator of outstanding panache, he promptly enlisted the two female staff to help us find a cheap hotel and guide us through the sites of the local area. They giggled as he worked his magic.

"You are the most beautiful ladies I have ever seen in the Salon de Provence Tourist Office."

We checked the 'Meteo' weather forcast. Our ferry to Tunisia was booked for 1pm the next day and the tailwinds, if anything, were to be even stronger. Able to stop and spend the afternoon meandering through the cobbled streets, Jamie and I did a whirlwind tour of the medieval fortress while Iain searched out an internet cafe to email his girlfriend.

A couple of hours later, ignoring the gastronomic delights of the many fine French eateries, we set our sights on the only Balti house in the region and ordered enough curry and naan to see us through the next nine months. Relaxed and sated, the guys joked over the remnants of a Chicken Dopiaza and a couple of bottles of Côtes du Rhône.

Susie's Diary—May 2009—Marseilles, France

Aaaaaaaaaah. Smack. Umpfh. Silence. (Feet. Check. Legs. Check. Body. Check. Arm. Hmm. That's a bit bruised. Hand. Oo, smarts a little. Face. Oh dear. That's definitely blood. And a bit of tooth...)

"Suse, you might not want to use your face to stop next time."

I would love to claim that my first real crash had been a dramatic incident in which I swerved to avoid hitting an elderly lady. In reality, I was distracted by the sight of a patisserie and pedalled into Iain—at 1 mph—by a mini roundabout.

Somehow in the ensuing second and a half I failed not only to hit the brakes but also to remove my hands from the handlebars. The only thing left to stop the fall was therefore my chin. We had joked before the trip about the problems of women traveling in the Middle East and how I could always disguise myself as a man. I didn't actually think, however, that just before getting the boat to Tunisia I would pick up an injury that makes me look like I have a goatee!

Ferry to Tunisia

Day 14—1306.93km

We sailed to Tunis.

Well, at least Jamie and I did.

Iain had snuck off.

Although his girlfriend was supportive of his venture, he was close to home, had a free weekend and was required back home. 'Flash' had duly ridden to the ferry port, given us his bicycle and hopped in a cab to the airport. While we would hit the seven seas he would fly to England, spend a weekend in the countryside and re-join us before we headed out of Tunis.

I mused on the delights of young love while Jamie laughed at him for being pu*sy whipped.

"Do you reckon he's going to last?"

"What—on the trip or with his girlfriend?"

"Ha—either I guess!"

Though Iain had insisted it was just a one-off, we both began to wonder: If he was flying back home already, would he really make it round the world? There was nothing to be achieved by worrying though. Iain himself was unsure quite what would happen. So instead we patched up my face, headed for the boat and focused on the unwieldy logistics of getting his kit across the Mediterranean.

And so it was that, 13 days since London, 13 hundred kilometres and 13 cycling shoe related accidents later, Jamie and I once again found ourselves in the car line up for a ferry; only this time looking far more haggard, spitting out cracked teeth and curiously pushing a spare bicycle.

Tunisia

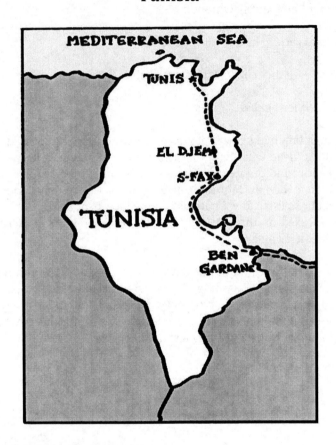

Estimated Distance: 563km
Actual Distance: 596km
Sunlight Hours per Year: Over 3000

Riding Tunes:

"Today is the greatest
Day I've ever known
Can't live for tomorrow
Tomorrow's much too long"

Today, Smashing Pumpkins

29 May—Tunisia

Day 15—1306.93km

Tunis was greener than expected and a deceptively easy introduction to the African continent. Not only did the local people speak French but, rather handily, our friend Mark was living there and could cycle Iain's bike back to his glass-walled apartment; his glass-walled apartment right on the turquoise ocean...

Mark worked for the African Development Bank and was not only able to introduce us to a host of fascinating people, but also to give us an insight into the continent's energy crisis. Which gives me the perfect—if decidedly unsubtle—excuse to add in a couple of words about SolarAid.

"In Africa, power is inaccessible, unaffordable and unreliable for most people. This traps people in poverty—students find it difficult to read after dark, clinics cannot refrigerate vaccines, and businesses have shorter operating hours.[4]"

In response to Africa's energy crisis the charity SolarAid was set up to get solar technology to some of the continent's most impoverished communities. There are 1.6 billion people in the world who have no access to electricity, yet many of these people live in the areas with the highest levels of sunshine. Through its social enterprise, SunnyMoney, SolarAid sells solar lights to these communities who have an abundance of sun; freeing them from a dependency on brutally dangerous, highly polluting and cripplingly expensive kerosene lamps.

If you want more details, you can head to: www.solar-aid.org

Just saying...

---

[4] The World Bank

Back to the story

On the overnight ferry we had slept and eaten and our time was to be filled wandering through medinas, ambling the Grecian old town and partying on rooftops. We were rested, prepared and—wardrobe aside—totally ready for it.

What we were not so ready for was the near-death kayaking incident...

Near death kayaking incident

In case of any doubt, I can, from first-hand experience, tell you that it is best to avoid kayaking to a nearby island when you see a storm coming.

In particular, it is best to avoid kayaking to a nearby island when you see a storm coming if the island you are heading for:

a) is not actually very nearby; and
b) if one of you is not technically in a kayak—rather holding on to one, lying on a surfboard.

Halfway to the island, in this particular instance, I think it is safe to say we all had the same thought.

"Why the **** are we still kayaking to a nearby island?!"

The increasingly emboldened wind had whipped the surf into an ambush and it was with weary resignation that we began to realise that the 'significantly further away than we thought' island was absolutely no closer than it had been 20 minutes ago. With no food, no water and no supplies, clearly there was only one course of action.

"Paddle harder!"

Afterwards, from the safety of his house, Mark remarked on the foolish approach to our decision making. He had wanted to turn round but didn't want to appear a killjoy in the face of two 'adventurers'. Jamie had wanted to turn round but didn't want to be the one to give up on the challenge. And I wanted to turn round as I quite needed a wee—but I didn't mention that either.

So we didn't turn round.

Some hours later, the waves now frothing in a white tipped frenzy, surfboard lost to the depths with Jamie clinging to the back of Mark's kayak, I was finally blown, arms screaming from the relentless labour, onto a desolate beach. Miles from the nearest town and in a scene ripped straight from the North African version of *Castaway*, the lone Tunisian fisherman walking there nearly keeled over when a blonde English woman washed up in a bikini and started babbling at him in broken French. This was not good. He was the only possible rescuer. With one teammate already the wrong side of the Mediterranean, it looked increasingly like the other was about to be hurled straight back across it.

The fisherman and I shared a moment of futile incomprehension before we both gave up hope and turned back to the sea. With no handily placed motorboat, neither of us could do anything but watch Jamie disappear from view as both he and Mark were smashed towards the boulders jutting their fingers into the turbulent ocean. It was a few heart-stopping moments of blind panic yet before both kayak and crew finally rolled through the surf.

Catching his breath Mark turned to Jamie:

"Thank God you let go before I hit the rocks."

"I didn't! I got cramp and fell off..."

With the last of our energy we trudged up the cliff to some local teenagers drinking whisky away from prying eyes. Persuading them to drive to help, Mark was wheel spun into the distance while Jamie and I lugged the kayaks up the cliff in the wake of the car's disappearing taillights. Where the road broadened we sat huddled on a rock, icy gusts washing in from the sea and slicing through our sodden T-shirts. An elderly couple drove past towards a village on the top of the headland, buildings outlined in the purple moonlight. They waved cheerfully to us and we mustered a wan smile in response. The vehicle trundled to a halt, its door swung open and the old lady picked her way towards us in the twilight. Taking the shawl from her back she muttered soft words of concern before silently draping it across our shoulders.

It was our first introduction to the kindness of the Tunisian people.

Iain called just as we were rescued and safely ensconced on the journey home. Calm and relaxed, he was waiting for us in a bar round the corner from Mark's apartment.

"Hey guys—what kept you?"

1 June 2009—La Marsa, Tunisia

Day 18—1320.25km

The next morning, more exhausted from the sea than from any days pedalling, we reluctantly left Mark's hospitality and got back on the bikes. The sun beat us ferociously as we weaved and dodged our way through the early morning rush hour and the concrete suburbs of Tunisia's capital. Cars and trucks spewed out thick black jets of smoke and pushed us perilously close to metal barricades, while oncoming vehicles jammed on their air horns and screeched to a halt as they slammed on their brakes. It made for a dispiriting ride.

On the edge of town Jamie got a puncture. His infuriation boiled over and his expletives grew coarser in direct correlation to the rising temperature, until a passing motorcyclist stopped to offer us a foot pump and we limped to respite in a nearby cafe. Within moments we were given cold drinks, handed sweet biscuits and offered our fill of succulent fresh melon before the owner presented me, rather obscurely, with a small pink plastic dinosaur. Fed and watered we joked with our genial host, once more the basic necessities of life restoring our spirits. I recalled advice from a friend of mine before we had set off: "If it gets bad, drink. If it's still bad, eat. If that hasn't fixed it, go to sleep and it'll be fine in the morning."

We named the plastic dinosaur Lance, in homage to Mr Armstrong, and fixed him to the front of my handlebars.

North Africa

Tunisia was very different from France.

Where we had once been riding alongside verdant fields and vineyards, accompanied by petulant skies and ignored on the empty streets of the towns we passed, now the landscape was dry and barren, the buildings ramshackle and people spilled from doors and alleyways. Women waved, motorbikes rode alongside us and children ran behind. The roads were lined with square concrete buildings, all inscribed with fading Arabic script. Those that sold food were made obvious by the barbecues burning outside and the carcasses hanging in front, shrouded with flies.

Lunch was obtained with a great deal of pointing and a barrage of smiling—the menu being largely what we could see and anything else the restaurant owner saw fit to bring us. It was hearty, basic fare of meat, rice and flat bread. In mine and Jamie's case, accompanied by plates of rudely chopped tomatoes and bitter onion. Iain had been stung before and insisted that scurvy could not be as bad as a riding with a dodgy stomach. In order to mitigate the chances we incessantly doused ourselves in antiseptic hand wash and scoured our insides with bottles of luminous orange Fanta. Fizzy drinks became a fixation. Our bodies had already begun to crave sweet liquids to replace the sugars lost by a morning of sweating. And it was only going to get hotter.

100 miles later, dusk heralded the call to prayer as we reached the city of Sousse.

Still too fresh off the boat to brave camping outside the city, we opted instead for a cheap hotel in its bustling Medina. Finding a gate in the Old Town's wall we pointed our bikes down the steep, narrow winding streets, dodged carts, swept past gem stores, swerved opulent fabrics and barely missed burning red spices, resplendent in their open-mouthed hessian baskets. Sending children flying and women shrieking, I bounded over the cobbles and careered to a halt inches before splitting a stall in half like a kamikaze two wheel ninja. It felt exhilarating. Like a pedal-powered Bond chase.

Surrounded by twinkling shards of light, we found a refuge in the heart of the medieval maze; basic but clean. Heaving bags, then bikes, up the steep stone staircase, we dragged all our kit to the tiny rooms above, each step aching through our protesting thighs.

March 2009—Somewhere on the M4, a few months earlier

I drew a deep breath as I wound up the window of the car and fumbled to turn off the radio. The last time I had been in the green bullet, Iain's Nissan Micra, we were on the way to Thorn, the long distance bike specialists, to order his tourer. This time we were flying down the M4 going through the presentation I had spent the previous week losing sleep over. While I would ramble on about solar nano-technology, Iain would wax lyrical about dreams, passion and exploration. Once again, we were pitifully under prepared.

Two weeks earlier we had been in Nokia's London office in a meeting arranged by Phil, who had run with me on the Marathon des Sables. We had sent through a proposal and outlined the aims of the project, hoping for sponsorship and some communications kit. The meeting had gone well. Nokia were behind the expedition and excited that our moveable solar power source would charge their phones. Using an inbuilt tracking application our journey could be pin-pointed, at anytime, anywhere in the world. They had agreed to support us on the spot.

We were amazed. It had been so quick, so easy.

A couple of days later an email came through asking us to give a motivational talk to the Board of Directors.

"A what Suse?"

"A motivational talk."

"Us?"

"Yeah."

"But we haven't gone yet. And Jamie's still in France. What did you say?"

"I said we'd love to."

I scrambled out of the car into the car park, abandoning my notes on the passenger seat and trying to keep my heart rate somewhere below 200bpm. Despite my nerves I couldn't help but feel excited. We were being randomly hurled into new and challenging experiences on a daily basis.

And I loved it.

2 June—Sousse, Tunisia
Day 19—1480.5km

In Sousse, my head sank into the comforting embrace of my pillow while the chatter of voices drifted in from the mosque outside. Adrenaline once again dissipating from me, I thought back to that moment in the Nokia boardroom. Since then so much had happened; each day a challenge, each more exhilarating than the one before. In just over two weeks we'd cycled through a country, reached North Africa and entered a whole new world of smoky souks and eye watering spices. Shattered, but contented, I dropped like stone into exhausted oblivion.

Too few hours later the dissonant wail of the dawn summons to prayer flooded through the narrow streets and torpedoed my peaceful slumber. True, the Medina was a fascinating melting pot of culture and tradition, but at 6am excitement about its beauty was somewhat tempered by the wake-up call.

Jamie refused to move.

He was an evening person, which was posing an increasing problem. The high temperatures and heavy mileage meant that an early start was imperative if we were to hit our target distances. But sleep was equally as important. Though increasingly accustomed to the rigours of riding back-to-back hundred mile days, we were nevertheless knackered at the end of every one. Iain and I had adopted the brilliant technique of 'falling asleep as fast as is humanly possible,' but Jamie just wasn't able to make himself nod off at 9.30pm. The weariness was beginning to tell.

Mostly, this morning, by his steadfast refusal to get out of bed.

"I'll get up when you get up."

"We're up."

"How up? On a scale of 1 to 10..."

Iain's Diary—May 2009

"The people of Tunisia have surpassed our expectations. They are incredibly kind, charming people who love to wave. The kids like nothing better than high-fiving you as you ride past which has nearly led to at least one high speed, embarrassing pile up.

They are also very proud of Hannibal, who rampaged through Europe with his elephants and gave the Romans a run for their money from Carthage, now a well-heeled suburb of Tunis. Long after Hannibal had stopped causing trouble the Romans had a major influence on this part of the world and the coliseum at El Djem is amazing. It's the third largest left standing anywhere and in its day would seat 30,000 blood-thirsty Romans watching scores of gladiators, animals and Christians meet their maker. When we arrived there were probably only 5 other tourists there which made the occasion pretty special."

El Djem, Tunisia

It had taken a while to rouse Jamie and a while longer to navigate the winding alleyways, but soon enough we were up, out and back on the road. And soon enough, too, the heat was burning.

Having read about El Djem, this magnificent Roman structure, we had decided to leave the sanctuary of the coastal road and turn inland.

Error.

Away from the sea breeze the air was stiflingly hot and resolutely against us. A commanding presence on the distant horizon, we saw the coliseum from 20 kilometres away, but even wandering thoughts of lions and centurions could not sustain the enthusiasm required for the interminable hour and a half it took to reach its gates.

We collapsed in the shade of a nearby cafe, and debated how long we could afford to luxuriate in the cool before heading back out to the inferno. A couple of local merchants eating at the table next to us regaled us with the history of this once fertile area; underground tunnels ran from El Djem's depths to the sea 40km away, enabling fresh supplies to be brought in when it was under siege. The image of these cool sanctuaries eventually lured us from our hiding place and we spent a blissfully peaceful hour pottering amongst the dusty ruins and taking stupid photos of one another. I stopped on top of the tall stone walls, soaking in the atmosphere and keeping one eye on the security guard rocking perilously back and forth on his chair while he slumbered.

In the distance I saw the first clouds appear, quickly followed by the ominous crackle of thunder.

A storm.

Where the rains in France had dampened our spirits, this torrential downpour, splashing into our faces and soaking us to the skin, woke us from our sloth and spurred us back onto the road. We raced towards Sfax with a renewed vigour and managed to reach the city just before dusk. Which was a great relief considering the slalom that lay ahead.

Never before, or since, have I seen such carnage!

Bullocks, ignorant to the traffic, pulled out from side streets, pedestrians meandered alongside trucks, and deep gouges in the road forced cars and motorbikes alike into abrupt loggerheads. We swerved, dodged, weaved and swallowed expletives in the face of the indifferent locals we narrowly avoided. We strapped Lance and one of the phones to the front of my bike making 'Lance Cam' with which to record the melee. It catapulted into a fruit stall as I slammed the brakes on at the crush of the first intersection. Relegating both to my pocket, I turned the volume up on my IPod as high as I could bear and re-entered the fray. Throwing my body into each sharp movement, the perilous enterprise became a real life adrenaline fuelled computer game. All accompanied by a cacophony of car horns, crashing metal and the pounding beat of 80's rock.

I was invincible.

3 June—Sfax, Tunisia

The guys were sharing a room that night.

When we weren't camping, we took it in turns. Sometimes Iain was my 'husband'; the next day Jamie. Finally it had come round for me to get some space. Once again we lugged our bikes, kit and exhausted bodies up steep narrow steps to a hostel in the heart of the Medina. This time though, my room was spacious, exquisitely tiled and backed onto a small patio with a view of minarets and the darkening horizon. I took a long hot shower and spread my few possessions liberally around the table-tops. Scoured raw, I eventually went in search of the guys, calling their names through the echoing corridors and locating them towards the back of the Tunisian maze. They were in a tiny cupboard, sharing an even smaller bed.

"What's your room like Suse?"

"Much like this...."

The next morning, Jamie's eyes bulged like a stress ball at bursting point and even Iain, excruciatingly cheery in the early hours, looked ragged around the edges. Though I had conceded the swap, the guys had been either too lazy or too gentlemanly to accept. It is fair to say at 6.30am they were regretting their decision. While I had slept the deepest and most revitalizing sleep in weeks, they had not fared so well in their sweatbox. We could not afford to linger in Sfax though. We had our sights set on Gabes, where we could take the day off to meander through the *Star Wars* film set of 'Tatooine' and the picturesque hillside region of Matmata.

Or so we thought.

'The Libyans'—as Monty Python would decry—had other plans.

In 2009 it was notoriously tricky to get into Libya. You couldn't get in unless you had a tour. You couldn't get in unless your passport was translated into Arabic. You couldn't get in if you were not in a group of four. And as of the 1st June—and they did not tell anyone until the 2nd June—you couldn't get in without a pre-arranged visa.

We had cycled 1,115 miles.

Three days from Libya, we could no longer get into the country.

Susie's Diary—June 2006

Cue: Panic.

Calls were made to the British Embassy, the Libyan Embassy and the Libyan border police. Mark, our friend in Tunis, even went to the Libyan Embassy on our behalf (once again, his generosity unparalleled). It wasn't looking good. Emergency plans were hatched.

"We could hire a fishing boat and sail to Jordan..." "Why don't we head north to Italy and work round the Med that way...?" "I could cling to someone's leg wailing until they let us in..."

Before a solution finally emerged. We could pick up visas at the Libyan consulate in Sfax, the town we had just left. There was no possibility of cycling back to get them in time. The train left in 10 minutes.

RUN!!

So yesterday we threw ourselves on the 4.10 to Sfax, had dinner in a brothel—cunningly disguised as a restaurant—and this morning, sought the cool sanctuary of the Libyan consulate where, after several heart-stopping minutes, we got our visas.

Libya is back on.

5 June 2009—Sfax to Gabes, Tunisia (again)

It was alarming how quickly we covered the same ground on the train. The blistering temperatures and constant toil of the road to Gabes were now replaced with slow gliding movement, the gentle ker-chunk of the track and quite a lot of biscuit-eating. The sun was low in the sky when we finally pulled into the station and, having narrowly avoided our first real obstacle, we decided to relax rather than risk a rush to Matmata before nightfall. The troglodyte cave dwellings of the *Star Wars* film set would have to wait.

Unlike the hectic activity of Sfax, Gabes had a slow and peaceful air. We turned out of the train station and sauntered down the main street. Jamie suggested locating a beer and, following the pandemonium of the past two days, Iain and I readily agreed. The only problem was finding one.

In the UK it is perfectly normal to nip to the pub or to pick up a bottle of wine for dinner. Coming from a country where alcohol is disguised as lemonade to encourage pre-pubescent consumption, it hadn't even crossed our minds to consider how complicated it is to get a drink in a society that frowns upon it. If you want a beer in the dusty, idle town of Gabes, for example, you get a briefcase, look slightly shifty and go to a heavily fortified building to pick up a couple of cans, hoping that none of your neighbours sees.

We saw the hubbub and went to investigate. Never before has it felt so illicit to buy a beer in a country where it is legal to do so. We, unabashed at our wanton habit, pushed our way into the small, crammed shop and placed our order. Several men, eyes to the floor, stood to the sides and, once they had succeeded in getting their paper wrapped wares from beneath a metal grill, immediately concealed them in the waist band of their wide legged trousers and scurried from the building.

We flounced out, loudly discussing the unusual scene.

Bloody tourists.

Susie's Diary—June 2009

Now a day behind—and without walking the hallowed ground of Obi Wan Kenobi—we were hot footing it to the border. Back inland...

Along the coast we are bolstered by a breeze from the sea. Inland there is also wind, just wind that comes a long way, from the middle of the desert. I hate headwind. It feels like going uphill all day. Iain hates too many hours in the saddle but we hadn't, until now, found Jamie's weakness.

Melting.

6 June 2009—Gabes, Tunisia

Day 23—1748.95km

To say it was hot the next day wouldn't just do an injustice; it would be like being caught at the scene, chased down the highway and incriminated by DNA. The days before had been hot. Very hot in fact—temperatures routinely knocking 40 degrees in the shade. But there is nothing that quite says "hot" like the combination of the battering summer sun and crippling desert winds of the Sahara.

My T-shirt had been glued to my back since daybreak and I paused to wipe my forehead on my sleeve, blinded by the stinging salt coursing into my eyes. I didn't care. It had been several hours since I gave up the battle with perspiration. Usually the guys would tease me as I insisted it was a lady-like 'glow'. Today they didn't even have the energy. We ploughed onwards in silence, the air thick and dense with the putrid smell of decomposing camels. There was no shade from the remorseless sunshine. Usually the movement of the bike created a cooling flow of air. Now, it was engulfing us in searing waves of heat and, as the gusts grew ever stronger, picking up sand from the cracked dry earth and flinging it at us, lacerating any parts of exposed flesh.

The landscape was desolate. The only cover from the stinging dust was the occasional shack selling dirty bottles of petrol or fly-strewn carcasses. Lorries thundered past our shoulders on the narrow, potholed highway to the border, sucking us into the vortex of their slip-streams and buffeting us perilously close to the vehicles behind. The only distraction was the occasional pair of cars pulled to the side of the road, their occupants in deep and animated discussion. I didn't look too closely; sure I would witness something less than salubrious. It may have been the worst day's riding yet, but getting a bullet in the head was still not a preferable outcome.

Full of gritty determination we rode on. I was suffering. Iain was suffering. But Jamie was falling apart. You could see it. Whereas before he would cycle ahead, full of energy and power, now he lagged behind. We would take it in turns to lead our fetid band so that those behind the front rider would benefit from their slipstream. In light winds we would go for 10km then rotate, in strong winds, 5km. Jamie's sections were notoriously rest-free. The pace would increase to such a rate that neither I nor Iain would get time to relax. Iain would take a more sedate speed in the lead and I would, irritatingly, change tempo randomly in line with my music. Today Jamie was the slowest—covered in a complete sheen of perspiration and looking shattered.

Visibility was diminishing rapidly in the sandstorm as we finally collapsed for lunch at a restaurant outside the town of Medanine. I went to the bathroom and stuck as much as my body as possible under the tap. Modesty should have told me not to wet my T-shirt in this conservative region but I was drenched anyway. We ordered food, sat and sweated. Even in the ambient temperature our bodies were unable to regulate. We were, to put it mildly, a mess.

This was the first real struggle we'd had. Until now we had turned in some pretty good distances, had a few climbs and the odd windstorm, but nothing that had really pushed us to our limits. Jamie wanted to stop in Medanine. Iain would probably have relented, but I wouldn't let them. Looking back, I am not sure what kept driving me and what kept me pushing the guys. Jamie was on his knees. Being a proud man, there is no way that he would have asked to wait out the heat of the day if he hadn't been suffering immensely. But we had a lot of miles to cover before we reached Ben Guardine and we had to make the border the next morning or we wouldn't get into Libya. There was no possibility of rest. We were getting back on the bikes and we were riding.

It was another five long and tortuous hours before respite finally came. As we rounded the corner to Ben Guardine I whimpered slightly as yet another lorry flung me off the side of the road into the gravel, my wheels slipping as I fought to keep balance. Exhausted to the core, I blinked back the tears of frustration and panic, flushed with a renewed vigour. Finally, we had made it.

Susie's Diary—June 2009

Libya is a harsh desert nation, flanked by the startling blue of the Mediterranean. We are following the coast road, skirting the sandy expanses of the Northern Sahara. Arriving here only three weeks into our nine month round-the-world cycle ride, we knew little more about the country than news headlines: the Lockerbie disaster and the strange habits of its mystifying and narcissistic dictator, Colonel Gaddafi.

The country, isolated from the world community since the 1980's, is still at pains to keep its personal business from the West. For this reason we have been forced to have a guide escorting us for the two weeks it will take us to traverse the breadth of the nation.

Which is how we met Lamin.

After a humid night being bitten alive in one of Ben Guardine's dirtiest motels, we awoke at dawn to make the most of the morning cool. Iain was first up, mainlining coffee to keep his eyes open. Jamie, as usual, was last. Though the most energetic of the trio he is still only a begrudging fan of the early hours.

The few miles from the border town to Libya were a series of stop starts. Numerous roadblocks meant numerous passport inspections and a forced enthusiasm for the increasingly familiar interrogation that ensues. Having found out that Libyan visa regulations had changed unexpectedly two days earlier, it was with some trepidation that we finally pull up to immigration control.

Was the travel company helping us legitimate? Would we be allowed in the country? Who was the man in a tattered brown suit grabbing my ear and chasing me round the hall?*

We sat on a concrete bench and studied every man entering the bare open-ended hallway. Could that be Lamin? What about the one with the briefcase? The clipboard? With several hundred pounds already transferred to the Austrian bank account of Azjar Tours there was nothing we could do if he didn't turn up.

After an unnerving 20 minutes, a tall dashing figure in a well pressed shirt headed our way. I was suddenly fully conscious of my bedraggled appearance and the pinpricks of sweat fighting their way onto my forehead. He shook our hands and broke into a toothy grin.

"There have been some problems but let me talk to these men."

And so Lamin became our fourth team member.

*We later surmised that he was the Swine Flu medic with a thermometer. He gave up after a protracted 'Benny Hill style' chase, presumably assured of my fitness if nothing else.

Libya

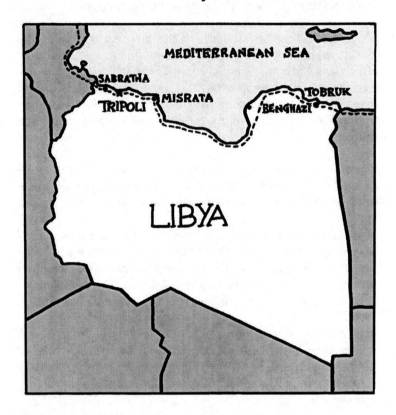

Estimated Distance: 1509km
Actual Distance: 1435km
Sunlight Hours per Year: Over 3000

Riding Tunes:

"Don't you know
They're talkin' bout a revolution
It sounds like a whisper
Don't you know
They're talkin' bout a revolution"

Talkin' bout a revolution, Tracey Chapman

7 June 2009—Libyan Border

It started awkwardly, with us never having ridden with a support vehicle, and Lamin never having driven quite so slowly down a busy highway—cars honking and truck drivers using the universal language of angry gesticulation. Gone was the open friendly welcome, children running alongside us, old men smiling and stall holders waving behind vast bowls of olives. Instead there were huge refineries on the horizon, terrifying lorries looming over us, a vast network of pipes, and the cheapest petrol in the world at 5 pence a litre. We hit the town of Zuwarah. The streets teemed with men, not even a burqa-shrouded woman in sight. Despite my long T-shirt and trousers, I felt naked and exposed to their dark stares. The guys didn't fare much better. This was not an area that saw cyclists. Even the sight of men in shorts was a knock to the arch conservatism of the region. It was not a happy place.

Initially our plans had been to stay here after the long border crossing but the day was still young and the conditions good. Without question we all agreed to push on another 40km to Sabratha. Lamin looked startled at the abrupt change of plans but, polite and helpful as ever, took it in his stride. We sped on; every pedal away from the inhospitable wastes of the border region a blessed relief. The day was unusually cloudy, holding back the heat. The cooling coastal wind was broadly in our favour. We were making good time. Not least, as we no longer had our panniers...

Susie's Diary—June 2009

First things first, we should probably make a bit of a confession. Well, not a confession as such but point out that we have enlisted a bit of help.

Libya is not an easy country to get into. As our visa situation so ably highlighted, rules and regulations can change on a whim and unorganised travel is not permitted. So, we have a guide.

And our guide has a car.

This automatically makes ours a less carbon friendly trip for which I will be off-setting. But to limit our impact on the environment (go with me here on the world's most pitiful excuse) we have decided that we will eat less if we carry less kit and have—just for this stage in which we have been legally forced to have support—put our panniers in the car.

Ok, there's a car driving 20kmph next to you... you wouldn't?

Really?

7 June 2009—Sabratha, Libya
Day 25—2040.32km

Without the 40kg of kit to weigh us down, we made it to Sabratha in the late afternoon—just as the sun finally broke through and pirouetted on the pristine blue of the Mediterranean. Sabratha is an ancient Roman city built on the trade of animals and ivory in around 170 AD, when Marcus Aurelius led the Roman Empire. We could see the tops of crumbling buildings as we rolled up to the lone souvenir stand in the empty car park and bought more neon Fanta to quench our thirst. The site would still be open for another hour. We began to lock our bikes and Lamin let out a laugh.

"You don't need to lock your bikes. You can leave them with my friend."

Jamie looked edgy and Iain hesitated. Everything we owned was on those bikes. Those bikes were our homes for the next eight months.

"Brilliant, that would be great." I interjected.

We had no reason not to trust Lamin, or the gentle old man proffering post cards and melting ice-lollies. This was not Zuwarah. It was a beautiful, untroubled place, steeped in majesty. If we couldn't trust our guide with three very cumbersome bits of kit, this was going to be two very long weeks.

We paid a nominal entrance fee and entered the site.

Unbelievable.

In 500 BC, Sabratha's port was established as a Phoenician trading post, reaching its peak under Roman rule. The now arid area was once a fertile gateway to Africa for rulers across the Mediterranean. They made great settlements along the northern coast of Africa at Carthage, Sabratha, Leptis Magna and Alexandria, building magnificent forums, temples and cities. In Sabratha much still remains—most astoundingly, the massive three-storey amphitheatre rising high above the water. In Rome, such a building would be thronged with tourists bumping into one another as they followed their multilingual audio guides or shouting tour leaders wielding umbrellas. Here, it was the security guards and us.

Iain and I sat on the wide stone steps while Jamie cartwheeled across the wooden stage. We breathed deeply in the gentle warmth of the late afternoon sunshine, imagining theatre once played out upon the dusty stage. We had made it here, to one of the world's most breath taking sites. We had made it through the heat and sand storms of the Tunisian desert; we had won the race for visas and endured a dispiriting couple of days in the flea ridden, desolate expanses of the border.

What a difference a day makes.

Jamie went to explore, Iain mused on the incredible advances of the Roman Empire, and I surreptitiously attempted to take artistic photographs of the men in uniform. After an hour or so our time was up. Though Lamin had gone to chat to the guards at the gates, our bikes—unattended—were just where we'd left them.

That night it was abundantly clear that things had taken a turn for the worse since the time of the Emperors. Lamin had done his best to find us some last minute accommodation, and though there wasn't a lot in town, we were cheered by the thought of a night in the government-run youth hostel.

Cheered, that is, right up to the moment we opened the door...

Now, I have stayed in some pretty bad places in my time. Places with bed bugs, grime woven into the sheets and overflowing toilets. Yet nothing quite compares to that youth hostel in Sabratha. A month earlier there had been a big conference of African heads of state, hosted by Colonel Gaddafi. The leadership had been wined, dined and pampered, while some of the less opulent hotels in the area had been commandeered for their security staff. Twelve to a room. Twelve men. Twelve men who had not only managed to back up the plumbing with their pubic hair, but also used part of the room to cut up meat, much of which had been left, festering, for a month in the Libyan summer temperatures.

Rancid hunks were besieged by flies that vied for air space with the smell and a biblical-sized scourge of mosquitos. Crumbs and peanut shells, animated by any manner of load bearing insects, littered the beds and were strewn liberally across the floor and—as a final flourish—there was an unfeasible amount of faeces splattered up the side of the toilet.

The cleaners, who had not been paid for months, had understandably taken one look and gone on strike. I was in no way joking when I suggested we should probably pitch our tents in the middle of the floor. I was sharing with Iain, who valiantly offered to take the first shower in order to clean it marginally. Jamie lamented the fact that his turn to have a room to himself was in such squalor. Sensibly, Lamin slept in the car.

After a restless night being dive-bombed and crawled over, we were all up and more than ready to leave by 6am.

What an amazing ride!

Well, Jamie and I thought so. Iain spent the next few hours wetting his pants in fear.

Part of what allows you to pedal around the world is a complete and utter ability to ignore imminent danger. If I had, at any point, really considered the safety of our endeavour I would have stopped, jumped in a cab to the airport and taken the first plane home. Not that we went out of our way to court danger, of course, but blotting out the potential consequences of our actions was a prerequisite that led to our new found favourite activity: dodging doom.

Like the ride into Sfax, for the 78km between Sabratha and Tripoli, it was a case of 'turn the music up and ride for your life', only this time with about 200 times more traffic. In fairness to Iain, even Jamie and I baulked slightly at crossing six lanes of a busy motorway but with Lamin guiding us up ahead, we eventually made it through the adrenaline-fueled ordeal and down the back streets of the capital to a hotel that was everything Sabratha was not.

Namely, sanitary.

Cultural differences

Normally in life, it is women who have those awkward moments. The bloke who stands too close. Wandering hands on a crowded bus touching you by 'accident'. A guy who doesn't take no for an answer. What a beautiful turnaround it is to watch a grown man squirm. Ah yes, ladies and gents, what I knew, but neither Jamie or Iain had yet realised, is that in many Muslim countries men hold hands.

Now, Lamin, our guide, had by this time come out of his shell a little. On the night in Sabratha we had been to the beach for a temporary reprieve from the hostel. Iain had stripped down to his pants and jumped in the sea. I had taken one look at the ladies fully clad in the black chador entering the water up to their knees and decided it might not be best to follow suit. Lamin had joked with Iain about his inhibitions. Iain had joked back. There had been quite a lot of joking. The mood was light. The guys were beginning to get on and take the mickey out of one another. We were starting to realise that, at 28, Lamin was a) even more immature than we were and b) very funny. What Iain had not been prepared for, however, was the moment when Lamin would reach across and take his hand.

If MasterCard wanted to sum up a look as priceless, I have the very one seared into my retina.

Fortunately for Iain his fears were alleviated almost immediately. Though he might be partial to a bit of man on man palm action, Lamin was also a devout adherent to his favourite hobby: fishing.

'Fishing', that is, for ladies.

The next two weeks were spent almost entirely in the pursuit of telephone numbers and a glimpse of rogue ankle. Even in the most obscure places, or perhaps only because they were obscure places, Lamin would find some young lady with whom he could pass the time of day. Even better at this sport was his friend Mohammed.

Early on, Lamin had come to us with a request. Since he had to drive behind us at 20km per hour for two weeks, could he bring a friend? And so 'The Big Black Man' joined the gang.

Mohammed wasn't all that big. Or all that black. But he was bigger and blacker than Lamin who had none of the qualms of racial stereotyping we avoid in the UK. Lamin proudly referred to himself as a 'cappuccino', regaling us with tales of Mohammed's dark skinned cousin, who couldn't be taken to the zoo in case someone thought she had escaped. Amazed that Mohammed found this hilarious, we agreed that one of us might need to have a quiet word with him if he ever came to visit us back home. In Mohammed's case it was not just the cultural differences that made life interesting, but also the lack of shared language. Most conversations with him were therefore either directed through Lamin or consisted of him punching Iain in a jocular fashion and declaring that he would soon go to the UK to steal his girlfriend. Repeat punch. Punch Jamie for good measure.

Much hilarity.

Play fight.

It was thus that I found myself on a road trip across Libya with four guys.

And, in the rather limited circumstances, naturally I quite fancied Mohammed.

9 June—Tripoli, Libya
Day 26—2114.02km

In Tripoli we took a day out to wash clothes, drink tiny cups of thick coffee and get our first taste of camel at lunch with a kind Norwegian man who stumbled across our blog and bought us a massive bag of peanuts. Libya is a dry country, which is conducive to early nights, peaceful sleep and early starts. We duly set off from Tripoli before dawn and benefited from not only the cool of the morning but a far less dramatic journey out of the capital than we'd had on the way in. Our rested legs made speedy progress of the coastal road and we were buoyed even further by thoughts of our destination. After the 130km ride to Al Khums, we would have yet another day off.

We had to, we couldn't miss Leptis Magna.

Leptis Magna is an astonishingly well preserved Roman city. Hidden under the sand for many years it was only excavated when the Italians invaded under Mussolini. Perched on the edge of the Mediterranean, the bath house, two forums and theatre preside magnificently over the far seas to Europe. It is, quite simply, one of the most astounding things I have ever seen. Striking. Unadulterated. Magical. And host to numerous images of phalluses that the tour guide pointed out time and time again with obvious glee. It was a fascinating insight into another world (the tour guide also took pains to explain the machinations of the two tier red-light district) and, once again, we had the ruins almost to ourselves. We dived into old bathing pools, clambered up archways and ran around the amphitheatre tw*tting each other with big sticks.

Eventually we returned to a delighted Lamin and Mohammed who, jaded from hundreds of tours of the ancient ruins, excitedly declared that the sight-seeing was over and that fishing would commence. We were off to the pharmacy; Jamie and Iain required lubricant.

In case you are not already privy to this captivating knowledge, male cyclists commonly smear petroleum jelly, Savlon or other such substances on their sweaty nether regions to keep everything in working order. Or, as I was once descriptively advised, to avoid 'boils the size of your boll*cks'. Lamin and Mohammed took to the task of purchasing some with aplomb:

1) because they found this notion highly entertaining, and
2) because the chemist is a place you routinely find women.

The guided tour of the world's most outstanding Roman ruins took three hours.

Buying lube for the guys' balls took nearly the same.

It was a hilarious escapade, not only due to Iain's mimicry of the lubricant's purpose but also because they eventually emerged, victorious... with an ointment for breastfeeding mothers.

Susie's Diary—June 2009

Libya is a big country. It is also a very hot country and—as we have subsequently found—an incredibly windy country.
It is 95% Sahara and has large tracts of 'not very much at all'. Much of which we have crossed and much of which we still have to.

The first day the wind Gods smiled upon us. We clocked up a grand total of 228km, by far our longest yet. It also helped that a brand spanking new—but as yet—unopened motorway was also on the route. We hadn't expected to be entertaining local buses with Superman poses as we flew through the Sahara but as we barely needed to pedal we could afford to mess around.

It was also immediately apparent to us just how lucky we were to have Lamin and Mohammed with us. Not having expected to get any help along our way, it was a strange feeling to be handed a cup of mint tea as we sheltered from sand storms behind the car and told that our dinner wouldn't be long. The guilty feeling about having support re-emerged but at least we had put in 144 miles to show willing.

The next day was again long and dusty—in fact—as have been the following four, so it all really begins to merge into one. What I can say though is that, when a wind blows across a barren expanse, you really want it to be blowing the right way. Twice yesterday I literally left the ground before being re-deposited. The very large, very fast lorries full of smiling faces are a joy to see in one direction as they 'turbo boost' you along but fill you with sheer dread when you see them hurtling the opposite way.

Other highlights include the joy that a random shower block by the beach can induce, the storytelling of our guides and the kind reception we have received along the way. Less fortunate is the proximity to traffic, the lack of sleep and the sheer volume of sand.

Which gets everywhere!

15 June—80km before Ajdabya, Libya
Day 32—2805.25km

The sun was still fairly high as we peeled off the road and pushed our bikes over the sand. Once again Lamin and Mohammed had scouted ahead and found a campsite, this time down a long covered track behind some dunes. We set up our tents with plenty of time before nightfall. We had a good fresh supply of water from the shower block the day before and Jamie hooked the solar shower over a nearby pylon so that he and Iain could wash. Hardly the most retiring, Iain immediately stripped naked and cheerfully soaked in full view, flashing his moonlike derriere and scaring the life out of our Libyan accomplices.

I contemplated washing under my sarong but opted instead for taking a bowl and hiding in a dip behind some spiky brush nearby. Even though there was no one to be seen, I washed bit by bit, revealing only my top or lower half at a time and doing my feet, legs and face separately to minimise the time I spent exposed. I then set about washing my thinning knickers and T-shirts in the part-used water to maintain a semblance of normality. Half an hour later I was back at the campsite and once again sitting on the day mat propped against the side of the car as the sun set and dinner boiled. By nine we were all in bed, resting up for the day ahead.

Voices.

Men's voices, speaking Arabic.

There was a flurry of hurried conversation.

I froze on top of my sleeping bag. Terrified.

The voices stopped.

Then started again. More animated. Sounding angry.

It was clearly an argument.

I couldn't tell how many people there were but it was certainly more than two. I prayed that Lamin and Mohammed would hear or were already involved in the altercation.

The voices stopped again.

The silence was deafening.

I could hear my breathing, every nerve ending in my body now tense and poised for action.

The zip on the outer sheet of my tent started to move.

Then stopped.

More voices. More disagreement.

The footsteps moved away.

Then returned.

I silently contemplated my options, heart hammering like a machine gun. I could call out, stay where I was, or see what was going on.

I patted around me in the darkness until I found my sarong, wriggled it onto my head and loudly unzipped the tent. There was no easy way of peering out of the small low entrance way and so I tumbled out gracelessly and staggered to my feet.

Three young men stood in front of me in the moonlight; jaws dropped as feverish eyes glanced at one another and then me.

I stared back. Unamused, but relieved they appeared unthreatening. I waved the back of my hands towards them gesturing for them to go away.

The oldest and bravest, around 17, took a step forward and pointed to me, then himself, then my tent.

I shook my head, slowly repeating the word 'no'.

He gesticulated again; this time with more enthusiasm and vigour, beaming as if he had stumbled across a quite marvellous idea.

I couldn't help but smile as I repeated my head shake and shooed them away. I felt nervous but now, at least, vaguely in control of the situation.

The 17 year old could not believe his luck. Taking my nervous laughter as acquiescence, he pointed once again at me again, then the tent, then his friends along with himself.

I laughed again.

There was clearly no way I was getting in the tent with him. There was even less chance that I was getting in the tent with the three of them!

I took a step towards them waving them away with a more obvious motion all the time shaking my head and saying 'no'.

The glances resumed and brows furrowed, my actions inferred that acquiescence may not be likely.

I took another step, shouting more loudly.

The youngest one took one look and ran. The others followed suit. But two steps out the oldest turned, rushed back and slapped me on the bum before legging it to the nearest sand dune.

I was shocked.

Not so much frightened as suddenly alone in the moonlight, not entirely sure what had just happened. I could hear the voices of the boys in the shadow of the sand dunes and heard a snore from Iain in the tent next to mine. I knew the guys where knackered, but I couldn't risk the visitors coming back. I crept over to where Lamin and Mohammed were sleeping and called out in a hushed whisper.

"Lamin." No reply.
"Lamin." No reply.
"Lamin." No reply.

I felt terrible. Hating to wake him up but knowing I had to.

"Lamin." No reply.
"Mohammed." No reply.
"Lamin!"[5]

A groggy voice responded and I heard motion from the tent.

Lamin's head appeared and I explained what had happened, laughing with relief and disbelief when I mentioned the slap to the a*se. Lamin looked shocked, then tried to suppress a childish giggle. He woke Mohammed and the two went in search of the boys while I was left alone again, now chilly in the desert night air. I saw the torches disappear into the distance and heard raised voices as they found their targets.

A few long minutes later Lamin and Mohammed returned, holding their sides, screeching with laughter. They too had assessed that the situation was not dangerous and were able to relax, explaining to the young men— alerted to our presence by Iain's flagrant cavorting—that I was not a travelling prostitute, but a tourist. And then, explaining what one of those was. They tried to look serious as they assured me they would not be back. Then broke into fits again.

"I cannot believe this boy, he hits you on your..."

---

[5] It later transpired that Lamin too had been frozen in fear that night. An attractive man, he confessed it was not uncommon for women on his tours to accost him as he slept. In one incident, so persistent was the lady he had been forced to feign impotence. He thought I was doing the same.

More giggling, Lamin doubled over with mirth.

"Shepherds. They know not how to fish!"

The next morning breakfast was full of excitement about events the night before. I was up early and discussing with Lamin how it could be confusing for young men, in a region where ladies cover their faces, to have a blonde woman pedal past flaunting elbows and ankles, when Iain wandered over.

"So, did something happen last night? I thought I heard a noise but had my earplugs in."

Lamin and Mohammed

To say we had landed on our feet to be escorted across the desert by Lamin and Mohammed would be an understatement. Though things had begun a little tense: How would we all get on? Would they try to change our plans? Would they get in the way? We had mentally circled one another, sniffed about a bit and found that we were all easy going, light-hearted and bound by a common desire to mess around.

Lamin and Mohammed were Tuaregs, an ancient people of the northern Sahara. Renowned for their storytelling, the tribesmen would tell us tall tales under the desert skies of witch doctors, smuggled 'fruits' and ancient Berber traditions. They would laugh and joke with us all, but look up guiltily if I stumbled across any conversations with Iain and Jamie that had turned to 'fishing' and German nudist camps. I was regarded as an innocent, sensitive female and dubbed 'the Queen'. All decisions were mine and in every argument I was right.

As it should be.

We had given Lamin and Mohammed one of the phones with a camera in it and they snapped away at us as they drove past pretending to be dogs, or jumped out at us in lay-bys. They would cook up a storm when we stopped, or flash their wide sparkling smiles at police checkpoints and in roadside cafes. They also spent a considerable time taking the mickey out of Iain, much to the delight of both myself and Jamie. Lamin would look across laconically as he whittled his teeth with a cleaning stick, gently dropping the mildest of acerbic comments.

"Iain, I looked at you the first time I saw you, and I knew you were a stupid man."

"Your girlfriend. She is beautiful. You. You are like an old tomato."

"In ancient Rome, Susie, she would be the Queen. Jamie, he would be an engineer and you, hmmm, you would clean the sh*t of the animals."

Brilliant.

It was all too soon that we had crossed the empty heart of the Libyan coastline and swung our bikes into the bustling city of Ajdabiya.

Susie's Diary—June 2009

"There are all sorts of things that we imagined could go wrong. Bikes breaking, illness, robbery—but the bubonic plague definitely came as a bit of a curve ball!"

16 June—Ajdabiya, Libya
Day 33—2984.05km

"What do you mean an illness?"

"My parents do not want me to go further because of the disease."

"But what disease?"

It took us quite a while to determine what Lamin was saying. Not only as he didn't know the English word for the plague, but as we simply could not conceive that our plans would be thwarted by the Black Death. The day before an outbreak had occurred in a small town near Tobruk, our destination on the way to the Egyptian border. Lamin, while not refusing outright to go, was looking decidedly unhappy about the prospect; fervently expressing a desire to take the coastal road and hop in a minibus around the infected area. With a dearth of alternative information and a strong desire not to receive a medieval illness we eventually opted on the side of caution and headed north.

If I am honest, I was quite relieved. East was a 350km road through the desert, north was straight into the Green Mountains. Well, 'greener' mountains. Not green by European standards, but with their proliferation of shrubs and glances of fresh air, they were still a blessed relief from the constant toil of the sandy expanses. We would not be able to pedal all the way to the Egyptian border as planned, but we would be able to pedal the same distance and end up at an ancient Grecian site into the bargain.

It didn't sound so bad.

Stopping in Ajdabiya for the day had not only given us chance to reroute, but also to check emails and maintain our contact with the outside world. I spent the time writing, blogging and sending messages about our solar, social and environmental findings to those back home:

1) Solar power—not much yet in Libya but with frequent power cuts in rural areas it would be a real benefit.
2) Refugees—many came into the country from Central Africa, some perishing in the sands. We passed several boats waiting to carry the desperate to Europe.
3) The sun—it was hot.
4) Very.

Jamie sensibly swerved the slow internet connection and set about checking bikes and kit and Iain got in touch with his girlfriend. All was not well. She had initially told him she could wait for the nine months he was away. Now she was not so sure.

There was a French upstairs neighbour.

He had proposed.

21 June—Al Bayda

Day 38—3338.46km

Al Bayda is the stop-off point for tourists visiting the ancient Grecian ruins at Cyrene. Much like the Roman site at Leptis Magna, this was an ancient trading city on the edge of the Mediterranean. This time, though, on a cliff edge and surrounded by burial chambers and masticating cows. The port was located at Marsa Susa, below the main site at the bottom of the hill. After another three days' pedalling we had plotted in another stop to clamber up to the overgrown temples and visit the ruins on the banks of the Mediterranean. Despite the large 'No Swimming' signs dotted about this archaeological site, some boys were splashing around in the water nearby. Under the baking temperatures, it just looked so inviting, so clear, so cool.

Jamie took one look, checked for security and without a care in the world jumped in and swam out sea. Naturally Iain saw no reason not to strip to his boxers and immediately follow suit. I was left on the bank sweltering and wishing desperately that I was able to do the same. With Lamin and Mohammed I waded in to my calves and watched with bitter envy as the guys splashed around in the luxurious water.

"Man, I wish I could jump in."

"Really? You want to go in the water...."

"Ahhhhh—you baaaaaaaaaaaaaaastar..."

Three seconds later the laughing guides had bundled me, fully clothed and screaming obscenities, straight into the surf. A minute after that, all now safely ensconced on shore, the armed men in uniforms arrived. They had heard the cries and run to see the source of the commotion. Like naughty school children we feigned innocence and tried not to snigger as my clothes hung off me in sodden incrimination. Two sets of unamused eyes slowly looked me up, looked me down then swept across the scene. Communally we held our breath until the soldiers finally turned in silent unison and walked away.

Perhaps I looked like a sweaty sort of woman?

22 June—Libyan / Egyptian border
Day 39—3338.46km

The next day was a flurry of bags, buses and border formalities. At the customs office the guards were nothing but kindness, but had clearly been issued instructions not let us out of their sight. Lamin advised us to hold tight, flash our most innocent smiles and ride out the inevitably long wait until we could cross to Egypt. The officer in charge of our papers offered me his chair with a flourish and we gratefully sat in the shade watching the dusty fracas of people trying to leave the country. Women, hardly able to walk under the weight of goods smuggled under their chadors, waddled to and fro. Buses with two metre stacks of luggage were laboriously checked, unpacked and reloaded. Arms flayed, baskets flew and in the midst of the organized chaos a man was dispatched to give me my first ever armed escort to the toilet.

Meanwhile Lamin had assumed his deferential 'I am talking to someone I don't like but need to be nice to' guise and did his best to negotiate our speedy exit. Freed without too much delay, we thanked our gentle captors and pushed our bikes through the gauntlet of angry faces lining the walk across no-man's land. This was frontier territory, trader territory—these guys meant business. With Lamin on one side and Mohammed on the other I picked my way through the glaring male faces; hostile tension bringing me back to earth with a bang.

Though not as much as the border fence did.

At the fence Lamin and Mohammed could come no further.

They had travelled with us as far as they were able and we could hardly imagine continuing without them. They had given us water in the desert, eased our way through each checkpoint and carried us across the sands with their quick wit and mischievous laughter. Without thinking, I propped my bike against a truck and hugged each one of them in turn. As I did, a disgruntled murmur rang through the watching crowd; eyes blazed, voices raised and yells rang out in displeasure. Alarm thudded through my chest at the surge of animosity but Lamin just broke into his signature grin.

"Susie—do not worry. They won't mind when I tell them you are the Queen!"

25 June 2011—Glastonbury, two years later

In the summer of 2011 I found myself staving off a seven-mojito-4.30am-winnebago-break-and-entry hangover in the BBC area of the muddy fields of Glastonbury. A few days earlier a researcher from Radio 4 had called and asked me to be a guest on a cycling-focused 'Excess Baggage' show, chaired by comedienne Sandi Toksvig. Having spent the previous 24 hours rigging sound stages, pitchforking hay and gate-crashing parties, I now sat entirely alone, head set on, in a newspaper strewn cupboard. Due to a festival-style mix up there was no time for a brief so it was with only seconds until airtime that I was shoved into position by a dead-eyed sound technician and the booming voices of the producer, host and fellow guests appeared in my ears. Unsure, even of my own name during the ensuing 30 minutes, all I can remember is blind panic and being jovially blamed for the onset of the Arab Spring.

Our ride had indeed taken us through Tunisia, Libya, Egypt, Jordan and Syria—all nations which, two years later, had seen civil uprising and a quest for liberation. Sadly none of my doing, it was with great joy for Lamin and Mohammed that I followed the demise of the Libyan regime.

The one and only time we publicly mentioned the name of Gaddafi, the self-styled 'King of Kings', Lamin's eyes widened in terror and he hissed at us to stop.

"Really? But we could be saying something nice."

"You do not know the risk."

Only under the desert skies did we talk of politics, even then in hushed undertones. It transpired that, after a fall out with the leadership, Mohammed's father had been forced to flee the country and that other members of his family had not been so fortunate. The country had been ruled by fear for the 41 years of Gaddafi's dictatorship. Over 20% of the population were government informants and dissenters overseas were assassinated. In 1982 Gaddafi declared that: "It is the Libyan people's responsibility to liquidate such scum who are distorting Libya's image abroad." He personally presided over the torture of those at home.

Since the revolution, Jamie, Iain and I have seen pictures of Lamin pop up on our Facebook pages with a host of other young men, peace signs to camera, joy writ wide across their faces. Much like the images emblazoned across the nightly news, they are waving flags, brandishing weapons and banging on the back of pickup trucks. In many, Lamin has a pistol or machine gun resting by his shoulder. It is a strange sight to see this most gentle of men at war, someone who squirted us with water, put cucumbers up Iain's nose and giggled like a school boy.

But it is a war that I am glad he has won.

Egypt

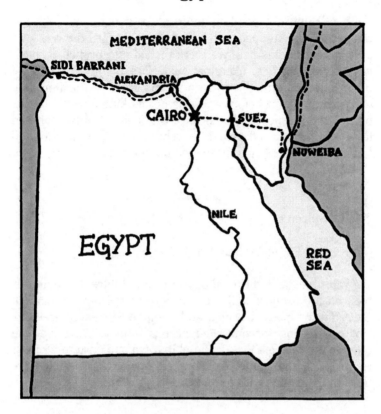

Estimated Distance: 1046km
Actual Distance: 1203km
Sunlight Hours per Year: Over 3000

Riding Tunes:

"I'm going in for the kill
I'm doing it for a thrill
Oh I'm hoping you'll understand
And not let go of my hand"

Going in for the Kill, La Roux

23 June 2009—Egypt

The scorching hot pandemonium of no-man's land subsided the second we stepped into Egypt. Ushered into the cool of a cavernous customs hall, we were picked and probed through yet another medical examination before being spat out the other side and straight down a steep cliffside to Sallum. Iain had been to Egypt before and was not looking forward to this part of the journey. Experience of the Sinai's tourist trail had left him convinced we'd be harangued, annoyed and immediately fleeced of our dearest possessions.

"Welcome!"

"Hello, welcome."

"Foreigner! Foreigner hello! Welcome to Egypt."

We were met with a cacophony of cries. Children on bullocks, men on motorbikes, ladies laden with bags from the market. Everyone we passed waved and shouted a friendly greeting. Not one asking for our time or our money. We pulled up outside the cleanest looking hotel in town and bartered for our rooms with the polite, well-spoken owner. It couldn't have been more at odds with our fears.

24 June—Sallum, Egypt
Day 41—3349.54km

Making the most of the cooler hours, we set off early again the next morning. Five miles out of town we got a tail.

We stopped.

They stopped.

We started.

They started.

The police van remained a discreet distance creeping behind, never getting closer or further away.

"Why do you think they're following us?" Iain was more curious than alarmed.

"Maybe they think we're spies?" I threw in, interest piqued. "Or they have secret military installations here? Or prisons. Maybe they think we're trying to spring someone?"

"Yeah—we could totally be hired assassins." Jamie joined in "Where would we hide a gun?"

"Easy guys, in the bike frame. It would be a brilliant disguise."

A detailed discussion ensued, revolving largely around the merits of the long-distance cyclist as a soldier of fortune, and the physical plausibility of concealing an anti-tank missile launcher in a hollow metal bike frame. Once again we were in the unoccupied desert, nothing around but miles of sand and the occasional white-washed building, each one surrounded by towering walls and sparsely decorated with rusty barbed wire.

Naturally, given that events had been running smoothly for at least 15 hours... we ran out of water.

In an act of abject stupidity, none of us had remembered to refill our bottles the night before.

Muppets!

If there is one thing even the novice traveller can probably fathom it is that riding a long distance in the midday sun of the Sahara might be best accompanied by a wealth of liquids.

Jamie was the first to get desperate.

"We need to break into one of these buildings."

"I hate to be the sensible one in any situation—but is that the best idea with the police 200 metres away?" Iain cautioned.

We pulled over and Jamie climbed through the dilapidated doorway of the desolate compound. Iain and I stayed outside, unusually despondent, and wishing for an inch of shade. Crestfallen and empty handed, Jamie returned. We pushed on, each now silent in our own meandering thoughts, dry air stripping the moisture from our parched mouths with every inward breath. All we could taste was dust and all we could feel were the pulsating sonic waves of heat undulating from the road ahead. Hardly in danger with a van load of policemen behind, my mind saw no reason to entertain such logic and descended gracefully into the panic-fuelled, heart-rate hastening sensation of mild hysteria.

It's a curious thing to be so thirsty.

It was another hour before we saw the tell-tale crates of empty bottles outside a shack ahead. Throwing our bikes down, we raced for the building across the dry, cracked earth and flew through its door-less entrance. Swerving the fly encrusted pail of water, we each grabbed a bottle of Sprite and desperately glugged down as much of the sticky liquid as we could manage. Gasping for breath and laughing in relief, chalky mouths were soon replaced with sugar-coated lips, wet tongues and clammy faces. The shopkeeper looked on in horror at the rapacious scene while the police shook their heads in dumbfounded bemusement. But wasn't until later that afternoon that the altercation with the authorities finally came.

On the edge of a busy trading town sat a simple, concrete, clean looking hotel. Iain took one look at it, one look behind him and veered over, trailed by Jamie, myself and a number of the local constabulary. As we padded our panniers into the lobby we were greeted expansively by the cordial owner who beamed in welcome and bounded round the reception desk.

Right until seven men in uniform trailed in behind.

There was much heated discussion.

We stood in silence, beads of sweat splashing gently onto the pristine marble floor, praying that the acrimony wouldn't end badly. Eventually an agreement was found and the owner sprang back into action.

We went to see the rooms.
The police came too.
We returned to reception.
The police lurked behind.
We heaved our kit upstairs.
They got settled on the couch in the lobby.
They watched us as we cleaned our bikes.
They watched us as we went for food.
They watched us as we went to bed.
And—at 5am—they got back in the van to follow us as we left.

That morning, though, we were determined to make a better start. Bleary eyed and groggy but now carrying our body weight in liquids, we had eaten a hearty breakfast and decided to befriend our escort.

They had AK47's.

We laughed and joked and tried to point out where we were going on the maps. A new police captain arrived. He was quite dashing. Hitting the road, I settled into a happy daydream about being swept of my feet by a dashing police captain.

The North Coast Road, Egypt

Never before have I seen sea the colour of that on the north coast of Egypt. Stunning, striking, vibrant blue—blue that you only ever see on postcards. Blue that literally takes your breath away, takes it on a sky dive and hangs out with it for a while at a Conservative party conference. Jamie had been reading a blog by a cyclist who had pedalled this route before and who insisted that, if there is one thing you should see in Egypt, it is the perfect blue sea of the country's north coast.

Sadly, our newfound cohorts had absolutely no idea where the coast road was and pointed us time and time again straight down the main drag, miles from its glistening shores. Assuming that they must know the area more than we did and having no idea what Jamie was talking about, Iain and I insisted on following the police instruction, much to his frustration. By the time we got to Marsa Matruh—and finally saw the water's staggering colour—he was ready to explode. Not only as we had now missed the most incredible section of our route but as, in town, the police continued to send us in the wrong direction. By this point Iain too was finding it hard to keep his temper. There was nothing for it.

PEDAL!

Cue: Keystone cop chase—hurtling across traffic islands, careering down passageways and hauling our bikes over dry stone walls; all complete with comedy back-tracking, hiding behind fruit stalls and the bellowed Egyptian equivalent of 'he's behind you'.

24 June—Marsa Matruh, Egypt
Day 41—3579.74km

"Red cobra to blue tiger, the Eagle has flown, repeat, the Eagle has flown." [6]

Naturally, by the following morning, the police had managed to track down three Western cyclists in a town of Egyptian holiday makers. We trudged down the steps of our hotel to be greeted by a brand new shift and a round-faced captain with fantastic English. Much of the previous confusion had been a result of failed communications and, after reprimanding us for

---

[6] *Susie's diary, June 2009*

'playing games', he provided us with an explanation for our unwanted escort.

Employing 12% of Egypt's workforce, tourism is vital for the country's economy. In 1997, Islamist fundamentalists, intent on Egypt's adoption of Sharia law, massacred 62 tourists at the Luxor Temple of Hatsheput. They killed 34 in the Sinai bombings of 2004 and 88 at Sharm el-Sheikh a year later. Hundreds more were injured. During 2006, 2008 and in February 2009, just before our trip, other bombings had taken the lives of dozens more and continued to strike fear into the hearts of those protecting the country's visitors. (Technically the police captain didn't quite express the reasons for our escort quite like Wikipedia does—but you get the drift.)

Randomly emerging across the Libyan border on two wheels, we had understandably created quite a stir amongst the constabulary of the northern coast and the local commander had insisted we were to be accompanied at all times. Finally understanding their concern and a little more grateful for our would-be protectors, it was a much happier entourage that set off on the 170km to Sidi Abd El Rahmen.

25 June—Sidi Abd el Rahmen, Egypt
Day 42—3750.24km

"If there is one thing I have learnt, it is that people tend to look at you if you rock up sweaty, dirty and riding a heavily laden bicycle. If you are sweaty, dirty, riding a heavily laden bicycle and flanked by six armed guards, you are basically a spectator sport." [7]

After a hilarious day messing around with the police, their car overheating from the laborious speed, we made it to Sidi Abd el Rahmen. Or at least to a junction in the road near Sidi Abd el Rahmen where several more police on motor bikes were waiting for us; Sidi Abd el Rahmen being nothing more than a

---

[7] *Susie's diary, June 2009*

series of beachside resorts which block the sea view from anyone wearily riding along the sand swept, breeze sheltered, tedious tarmac coastal road. Again, the authorities would not let us camp but diligently set about finding us somewhere to stay. Exhausted from a day of debilitating headwinds, our guards—come tour guides—were proving a godsend.

Or so we thought.

Another five miles later, as dusk was falling, we reached a deserted hotel and were shown to an apartment which would have been incredibly luxurious if it weren't covered in sand. Totally covered. Not just a bit covered. Covered like those pictures you see of old towns that have been swallowed up by dunes, long forgotten and lost forever. The owner, who had clearly been told to let us stay by a number of determined looking men bearing machine guns, diligently, if unhappily, set about removing the sand.

For the next four hours.

At around 11pm we were eventually let into what was still the sandiest room this side of a 'Biggest Sand Castles of the World' competition, and settled in for a few hours of general discomfort being mauled and sucked on by a squadron of merciless mosquitoes. Exhausted, with sunstroke and unable to sleep, it was not surprising, then, that Jamie picked up a fever.

"Do you think you can cycle Jamie?"

"Maybe. In an hour. Can I go back to sleep for a second?"

By 9am the next morning, Iain and I were getting impatient. Despite the fact that Jamie was clearly in trouble, we too had slept fitfully and the weariness was beginning to tell. As we waited, the thick blanket of heat crept its arms around us and dragged us further into despondency. We wanted Jamie to make up his mind before we were faced with cycling straight into the oppressive fire of the midday sun. Either he could ride or he couldn't. The police had arrived and were urging us to stay or decide that we were going to move. Jamie was burning up.

We all knew he shouldn't get on his bike but he was loath to accept it.

Alamein, Egypt

In 1942, amongst the stifling summer heat of the desert, a fierce battle began between the Allies and the Germans over North Africa and its pivotal supply routes. Three months and 40,000 casualties later, it ended in an Allied victory. On the road between Sidi Abd el Rahmen and Alexandria a well-tended graveyard stands on the parched, scorched earth of Alamein. We stopped to see the memorial. Iain and I ambled quietly through the rows of headstones, remembering those who fell. Few were older than 25. Jamie collapsed on a bench by the museum, too delirious to pay attention.

"We've got to get him to stop, right? He's killing himself."

"It's his decision Suse."

So many graves; so many young men. I was annoyed that Jamie was not being rational and admitting that he was too ill to continue. Firstly, as he was clearly not doing himself any favours, but secondly, as he was moaning about being sick in a place where thousands lay buried. We had chosen to pedal about in the midday sun. It was doubtful they had opted to wind up in a desolate corner of the planet blown to pieces by a Panzer.

"Guys. I have to stop. I keep blacking out."

"Thank God!"

Luckily today we knew exactly where to send him.

Months earlier, holed up in my Peckham kitchen on a rainy afternoon, I had fired up the computer and Googled 'Egypt' and 'surfing'. Not too hopeful about the waves in the Mediterranean, my optimism had nevertheless turned to sun, sea and splashing about in it and I was excited to find a website for the Egyptian Surf Riders Committee. Weeks later, after a mind blowingly frustrating 48 minutes in a Libyan internet café, I got back in touch with its founder, Tim. Tim was the co-owner of Adham resorts and had invited us to stay.

"Err—so Tim—you know how we're cycling round the world."

"Yup yup, amazing stuff."

"Um—thanks—yeah—well—Jamie's going to be arriving in a cab..."

We packed Jamie, still reluctant, in a van while Iain and I laboured on down the busy highway. We found Tim a few miles later with his wife, Gaela, and daughter, Nora. They negotiated our release from police protection and guided us back to the sanctuary of their hotel.

26 June—Alexandria, Egypt

Day 43—3864.24km

Pool, workshop, laundry, beer, food, bed.

Heaven.

We spent the next two days in indulgent lethargy, enjoying not only the rest but the generous chatter of fantastic company. Jamie recovered and so did all our spirits. We played tennis, drank wine and rested in the warmth of the Adham family. Through all of this, though, there was one huge and lurking issue.

Iain's girlfriend had pulled rank and he'd dropped the bombshell.

"I have to go home."

"We thought you might."

"Just for a bit. When I've sorted things out I'll come back and find you."

"You know you're always welcome but don't worry if you can't."

"She'll dump you anyway when she comes to her senses mate. Mohammed's probably moved in already..."

I tried to remain sensitive and understanding while Jamie joked around. I was gutted though. Iain was a constant source of entertainment. His ridiculous attempts to flirt with anyone we rode passed had us crying with laughter, his ability to make up facts about everything we saw kept us distracted from the heat and his knack of tracking down the most salubrious establishments was an unparalleled boon. He had also, without realising, been a great emotional support for me.

Having known him from the age of twelve, ours was an easy relationship, nothing more in it than a mutual fondness and a desire to cause mischief. He had always been the errant older brother and the worst kind of example; the one who'd keep an eye out for you but get you in more trouble than you even thought possible. He would joke and tease me, but all the time make sure I was happy and breathing.

Jamie was more of a man's man.

I am smiling as I write that. Thinking of the grief I'll get from Iain for it! But what I mean though is that, although perfectly capable and extremely good at talking about emotions, it was not something Jamie and I would do in the casual course of conversation. He had not grown up knowing tales of my mother and I had never been accosted by his, at 2am, sliding down a wall clutching an empty bottle of Liebfraumilch. Having ridden to Spain and been friends since university, Jamie and I were extremely close, but it was not yet the same as 18 years' worth of empathy.

For both of us, though, it was a blow to lose Iain; a third of the team—or half by his body weight.

When someone leaves there is a complete shift in dynamics, as Nial and Timmy's departures had already taught us. There is one less person to diffuse a situation, one less to share the chores, one less to do something stupid or to distract the local constabulary. With just two people the relationship can't be anything other than more intense. If one person's ill then the second does everything; if your teammate annoys you, then there's no one to turn to.

But cycling round the world is not something you do if your heart's not in it. It's not like going on a night out because your friend really wants you to. It's putting your whole life on hold, for months, and going to the far and distant corners of the globe—without air con or a seatbelt.

We couldn't be angry with Iain.

We'd just see how we managed without him.

Susie text to friend—June 2009

"Hey Meens!!! Tracked down some surfers in Egypt. Who'd have thought? No gossip. But—hold on—wait for it, wait for it. I did touch knees with a man. Yep. And he didn't move his knee. Ha. This is the closest I've got to romance. Don't get too excited! Anyway, everything ok? What's the news?"

Despite the 'knee incident' with a guy in the Adham bar, Iain's departure brought only one type of comment.

"Oh really? You and Jamie on your own hey..."

With Boris Johnson, Mayor of London, at City Hall.

"I don't get it, why've we stopped here... aah!" Prenois, France

Lunch? Tunisia

The Boys, Iain sporting Lamin's best 'fishing' outfit. Libya

Kuraymat Concentrating Solar Power (CSP) plant. Egypt

After a visit to the 'Putting on Special Clothes Room'. Syria

Oy! Which one of you slapped my ar*e? Syria

POW! Turkey

Demonstrating the GPS, Turkey

This family followed us for 25km, capturing the whole thing on candid camera. Forget the boys in the picture above, looks like Jamie has another fan, Iran

In the desert with Zhenya. Turkmenistan

Carrying Jamie's panniers I finally get a puncture. Dammit!
Turkmenistan

Finally able to relax. Beers with Zani. Uzbekistan

Iain makes it back to join us on the Silk Route, Samarkand

Unexpected traffic jam. Climbing into the Tian Shan mountain range. Kyrgyzstan

Worth every long mile and sand-covered hillside. Jamie near the Irkeshtam Pass, Kyrgyz / Chinese border

30 June—Cairo, Egypt

Chaperoned by the army for our ride down to Cairo, the first day was easy if overwhelmingly toxic. The petrol filled highway was as flat as a pancake and being pounded by lorries and blacked out estate cars. The truck load of policemen was no longer with us, but we'd been given a platoon to supplement our protection. Delighted with the upgrade and spurred on by the regiment, our only small setback was a multi-hole puncture. Checking the maps we had assessed that the distance was too much for one day, so we stopped mid-way to the city in a motorway guest house. It was clean and quiet, but my turn to struggle.

The temperature was unrelenting—thick, dense, airless weight of heat that stifled our lungs and wouldn't let go of us. I got up and had a shower, then stuck my clothes and pillow in the water too in a faint attempt to cool them. Nothing was helping. I opened the door. Thud. More heat. No relief, just an excitable onslaught of relentless mosquitoes. I burned through the night, swatting listlessly and dreaming of fresh air and the chill of the morning. It never came. We would often get up in the purple of daybreak, hitting the road in the moderate conditions, but today the dawn was already savage and compounded by the fear of a strengthening breeze. With no sign of the army we trudged our bikes back to the highway. More fumes. More heat. More ceaseless toil of hammering headwinds.

On the outskirts of Cairo we saw a shopping mall and made our way over, ever so slightly in awe at the sight of it. We bought ourselves drinks and a couple of sandwiches and were dutifully followed by the suspicious security guard.

"We must look pretty bad..."

We were 30km from the centre of the city and already we were two hours late.

At the Adham Resort we'd met Wesley and Steve, two guys working in Cairo and out for a day trip. An ironic Scot and a jocular American they would fly back and forth from the rigs on the coastline and were the generous inhabitants of company apartments. With extra rooms and benevolent natures they immediately bought us beers and invited us to stay. In the fuzzy glow of mild inebriation, Wesley peered at the screen of Jamie's solar powered Sat Nav and plotted in the name of our destination.

At 2.30pm we finally ploughed into the 20-million strong inferno. Plastic bags flew at us like tawdry parachutes, metal shards spiked from the roadside like daggers and old shoes and overcoats lay amongst rubbish tips. Dirty buildings, black with pollution, flanked the sides of the heaving overpass and we swerved from it down a slipway to the dusty streets below. We flew past tyre shops, skidded round man holes and came to a standstill outside some laundry strewn tenements. The red dot on the phone screen blinked our arrival. It did not look exactly as we might have expected.

We called Wesley.

"Can you see McDonalds? KFC?"

"No, but there is a chicken running up the street..."

Several children came to chat and were summarily removed by their anxious parents.

"Are you ok? Can we help?"

"No, no, our friend is picking us up."

"Really? Here? Are you sure?"

We were lost.

After much confusion and deliberation we got back on the road and headed south into the heaving pandemonium. Still, Wesley, to whom we remain eternally indebted, was unable to locate us.

"Can you see anything written in English?"

"Does the writing on our bikes count?"

The other side of the river I spotted an HSBC. We headed the wrong way up the motorway and careered across. I was directed to the Customer Care area and—with Wesley on the line—thrust my phone at the slightly frightened looking lady sheltering behind the counter.

"Help!"

The dirt, sweat and look of desperation on the face of someone who, suffering from sleep deprivation has just cycled backwards through Cairo in the midday sun, is probably not the most reassuring vision—even through bullet-proof glass. Kindly, though, she acquiesced. I can only imagine, as it seemed the quickest way to remove me from the building. After telling Wesley where we were she visibly relaxed to find there was an explanation for my state and directed us to a nearby hotel so we could wait for him.

The Marriott.

It's fair to say we looked out of place.

Wesley's apartment

In case you are ever using solar powered GPS to navigate the megalopolis that is modern Cairo and making your way to Bourg Said street, please note: there are two. One in the leafy westernised suburb of Maadi, and the other under a death defying, fast moving, multi-laned ring road.

Beer in hand and happily shrouded by the air conditioned temperatures of Wesley's apartment, we laughed at our misfortune. Jamie demanding an apology for the vitriol poured onto his directional abilities, and me offering him the broken plastic clock a child had given me, as a peace offering. For all its discomfort, our diversion to Cairo's northern suburbs had been a heart-warming experience. The next morning, however, warm-heart had been resoundingly replaced by stomach-in-throat.

"Blurgh! Jamie, this is not good!"

I groaned and curled up on the edge of my bed.

In ten minutes we would be picked up and taken to the Kuraymat power station. We had cycled thousands of kilometres, battled the Sahara and contended with everything from rampant goat herders to the bubonic plague. Now, the morning we got to visit our first actual solar power station, I was puking explosively into the bathroom sink.

My stomach, already feeling temperamental, had clearly not taken well to the previous evening's deluge of rich food and alcohol. I dragged myself up and into the car, looking paler than should be possible after two months in the desert sun and huddled on the back seat clutching onto my aching stomach. The driver looked delighted and rubbed his belly.

"Oh man. I thought it couldn't get worse. He thinks I'm pregnant!"

Kuraymat Concentrating Solar Power Station, Egypt

"Tucked away roughly 90km south of the bustling streets of Cairo lies Kuraymat, the site of Egypt's first hybrid solar power plant... Spanning across the uninhabited desert landscape, the plant will collect solar energy through a total mirror surface area of 130,000m². Kuraymat will feature parabolic trough technology integrated with combined cycle power using natural gas as a fuel. Combing the product of natural gas and solar absorption, the hybrid power plant will be capable of producing

150 megawatts of power, a solar share of 20 megawatts... The hybrid power plant can be operated round the clock and will serve 500,000 households with electricity."[8]

The hot air punched into me like an electric snowball as I opened the door of the car. I let out an unintentional moan and mentally assessed the chance of collapsing.

Moderate.

"Jamie you're going to have to do all the talking."

"Will you be ok? Of course, no problem."

What a dude.

We were greeted by the smiling faces of Hisham and Mohanad from Oracom who explained the complexities of the station with compelling levels of enthusiasm. The site had been chosen for its abundance of sun and close proximity to transport, materials and the waters of the Nile. The project was being backed by a number of international companies, agencies and the Egyptian government. It was a fascinating, progressive and—in my sensitive state—ridiculously bright, endeavour.

Thank God for sunglasses.

If you are heading to a place in the desert covered by thousands of mirrors, it is not ideal to arrive hung-over, with heat stroke and sporting an acute case of Pharaoh's Revenge.

Joking aside, my nauseous condition was a bitter disappointment.

Having got the go ahead to visit the site from Flagsol, the German company part funding the Kuraymat project, and following such a great reception from its managers, it was galling to be too ill to maximise on the experience. I had been

---

[8] *thedailynewsegypt.com*

in contact with the British Council and met a journalist at the Adham who had offered to get our story into the local papers. But as I convalesced in Wesley's apartment, the chance to promote the CSP plant and the aims of our trip had slowly slipped away. Blogging from the road was not proving as easy as I'd imagined, nor getting information back to the press at home.

If only I'd had a bit more willpower when that Texan oil magnate had handed me the fourth gin and tonic...

Sabotage!

5 July—Suez, Egypt
                                        Day 52—4249.79km

Imagine.

It is 6.30am. You are already covered in a sheen of perspiration. You are groggy from sleep and have just, unwillingly, shoved a leaden chunk of bread down your gullet. It sits strangely in your stomach as your dull head tries to wake up. Every pothole you hit reminds you that concentration is paramount, as does the passing of each ten ton lorry and every bus which cuts you up, time and time again, relinquishing its passengers. There are hills. It's smoggy. You are guzzling water with the niggling concern that you may not be able to replenish it. You are the slowest thing on a three lane motorway.

And so we left Cairo, the calm of the Felucca ride we had enjoyed the night before, already a distant memory. But it was off to Suez or Seuz/Suze/Sezu, depending entirely on the linguistic skills of the various sign writers, and then to the desolate interior of the Sinai Peninsula. Sinai is usually known as a tourist mecca. Snorkelling, diving and kite surfing abound. Naturally we were missing all this and heading straight into the desert.

Again.

6 July—Sinai Peninsula, Egypt

Beating sun, gale force winds and mountain passes all without food, water or shelter. The Sinai certainly decided to test our mettle. Luckily, the two times the next day that fears of dehydration wrestled into my consciousness, we stumbled across villages replete with military outposts and huts selling hot sticky cola. The guards delighted at the reprieve from their lonely station, poured over our maps and kit and broke the tedium of our thankless ride. In the second village, a donkey entertained us as by sauntering into the road and refusing to move in the face of a 20 ton juggernaut.

I took notes for next time I fought with my mother.

At the owner's insistence, we pitched our tents that night on the porch of a busy roadside restaurant, avoiding the threat of scorpions. Next to the all-night kitchen, we dozed fitfully in our airless shelters while being simultaneously violated by both the shouts of its proprietor and the excruciating shriek of Egyptian soap operas.

But though one day can bring an arduous ride, the next can fly.

After the densest coffee I have ever had the misfortune to choke on, we set off into the tranquil mist of a perfect desert dawn. Floating through the eerie fog, the water wrapped around us like a cooling shroud, thick drops turning into translucent crystals which danced on the frozen hairs on our arms. As the haze cleared, huge black sand mountains emerged either side of the road and stretched out beneath the horizon. The desolate beauty gave way to games of 'count the Israeli gun towers' and a final dramatic plummet to the coast. We had climbed (oh boy, had we climbed) but what comes up must, indeed, come down.

As the heat of the day began to drain us and dull our spirits we took a sharp turn to the left and headed straight into a canyon gouged through the imposing cliff edge.

High-speed downhill for about 20 minutes.

Just cruising.

Checking out the scenery.

All the way to the sea.

Jordan

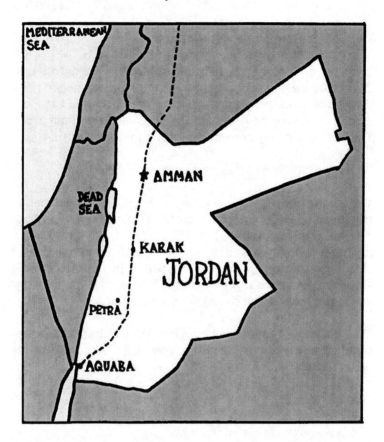

Estimated Distance: 364km
Actual Distance: 553km
Sunlight Hours per Year: Over 3000

Riding Tunes:

"It's been a long, a long time coming
but I know a change is gonna come, oh yes it will"

A Change is Gonna Come, Sam Cooke

8 July—Jordon

The next day it was all about the ferry.

If you want to enter Syria, Lebanon or Iran, you are not allowed an Israeli stamp in your passport. As our journey onwards was through all of these nations, we skirted the Promised Land and made for the ferry to Jordan. The scheduled two hour ferry. The ferry commonly known to locals as: the 'slow boat'.

"You are lucky. Sometimes it takes three days!"

Though the journey was a long eight hours, it was a fascinating trip holed up with two stoned Jordanian students and an Iraqi refugee who had risked her life to sail on a tiny boat to Australia many years before. She was not happy. Having steadfastly refused to holiday anywhere in the Middle East since her escape, she had eventually conceded to her new husband's request and come back to the region.

"The dialect of Arabic here is similar to that in Iraq. I started to wear a burqa again as I could hear what the men were saying when I didn't."

"Really? What?"

"Trust me, you don't want to know!"

Susie's Diary—July 2009

Passport confusion at the other end led to our first night cycle.

But it was not to be the last.

After a short night's sleep we left the Jordanian port town of Aqaba at 6.30 the next morning. Jordan is clean, beautiful and... somewhat mountainous. The plan was to cycle the 130k to the ancient city of Petra, one of the Seven Wonders of the World and destination for Indiana Jones on the Last Crusade. The road out of Aqaba heads straight uphill. Again we were pedalling into the wind. When we stopped to get water at 9am, we realised that, at the current rate, the day's ride would take a further 17 hours.

Events continued to conspire against us. The trudge was unrelenting. One of the funniest things I have ever read was Ben Elton's description of two old ladies having a Zimmer frame race. Progress was much like that.

To call it snail's pace would be generous.

We stopped for lunch at two, exhausted. We would normally be nearly finished by this time but had only gone 60km. We took a while to rest and to try filming some footage to send back home for the 'We Support Solar' parliamentary campaign. We were pleased to use our trip to help promote a feed-in tariff for solar energy in the UK. And equally happy to have an excuse to stop! Begrudgingly, we got back on the bikes.

17km later we hit a road block. A lorry had fallen over and was being towed off the hill. It took 45 minutes. It was nearly 6pm. We were still a long way from our destination.

Diary insert

It is at this juncture that I should add a tiny insert to my diary entry. There is a small tale of stupidity that I failed to mention at the time. Real stupidity. Omitted partly as I was ashamed of being so foolish and partly as my parents were anxious enough already.

It goes like this:

Hill after hill after hill.

Normally I could struggle on. Throw my mind somewhere else and ignore the situation. Keep churning through the miles and pushing the pedals. That day, I couldn't. Not couldn't for lack of motivation—though that was scarce—couldn't as I had lost the lowest gears on my bike. I had tried to fix them that morning but already exhausted from the ferry and another sleepless night of sweating, I soon gave up, hoping it wouldn't be an issue.

It was.

After several hours of up-hill pedalling my legs gave way. I simply couldn't turn the wheels. I got off to push on foot. A car pulled alongside me offering a lift. I declined and continued the vertical slog. A few minutes later another stopped. Once again I refused the offer but by the third, my resolve was wavering. How bad would it be? My bike was legitimately broken. We would never get there before nightfall. Jamie had missed a section of the route from illness, so technically I had already cycled further... I threw my bike in the back of the truck and jumped in, waving at Jamie as I sped past.

Five minutes and two miles later, we were at the top. The driver explained that he was an off duty policeman. He went to a shop by the side of a road, bought a melon and borrowed the stall-holders knife to cut it up. We nattered in the late afternoon sunshine. Jamie arrived and the policeman explained that there was only one more bit of steep up hill, pointing at the crest above. He could take me and we could all meet at the top and eat the melon. Jamie didn't look pleased but I readily agreed and off we drove.

Only, of course, we didn't stop at the crest of the hill.

Or around the next corner.

I started to get nervous and asked to pull over.

"It's ok, we're nearly there."

"Where?"

"To the place you get a view of my brother's farm."

"But I thought we were stopping at the top."

"It's just here."

"No we have to stop! Jamie will be worried."

"Just a few more metres..."

I was nervous, my voice starting to waver. I rationalised that nothing bad had happened and prayed the guy was genuine. I mentally assessed how much damage I would do to myself if I jumped out of the moving vehicle and how I could then get away without my bike, passport or any of my belongings.

In the end I said nothing but the jovial mood was suddenly tense and awkward. We drove on and on, for what seemed like hours, but could only have been a few minutes. Though eventually we did stop and we did look out at his brother's farm, we were now nowhere near the place we were meant to meet Jamie and nowhere near anything other than a sheer rock face and a smattering of mountain goats. It would be a while before Jamie got to us.

The driver tried to get a bit closer across the front seat. I got out of the car. He came round and started to tell me how beautiful I was. The situation was not good.

Back on the bike, all Jamie knew was that he had just seen me drive off with a man wielding a machete and I was now nowhere to be seen.

By the time he arrived, I was ready to run, and he looked shattered, anxious and on the verge of exploding.

I had already taken my bike out of the truck so literally hopped straight on it and began pedalling.

"Suse, if you do that again to me, forget anyone else killing you!"

Back to: Susie's Diary—July 2009

By nightfall we had climbed over 2000 metres and there was 30km to go. But we were determined. Right up until the dogs. And the cars. And the potholes...

The road to Petra is called the King's Highway. It is stunning and—in the Southern part at least—populated by Bedouin people. Bedouin people who keep sheep. And dogs to protect their sheep.

Despite the fact that my lights had broken and it was pitch black (Mum—would it help if I pointed out that I was still wearing a helmet?) the dogs heard us coming and gave chase. Though we kicked and screamed they viciously gnashed at our ankles, teeth bared, spittle flying in the moonlight. Twice we were saved by buses coming in the other direction. A blessing at the time, though traffic is not always a welcome sight when you are freewheeling, unlit down a pitch black mountainside.

We hit a small village and—auspiciously—the first car to come round the bend was a taxi big enough to take bicycles...

9 July—Petra, Jordan
Day 56—4683.2km

It was still 20km to Petra, the 6th century Nabataean city cut deep into the rocks of Wadi Musa, when we pulled into the small hillside village. Adrenaline giving way to utter annihilation, we took one look at each other, one look at the perfectly sized vehicle and took all of three seconds to jump straight in. We would get to a hotel and retrace our steps later. It would be safe, simple and we would finally relax.... our amiable driver had other ideas.

Naturally, on a day where we had seen a lorry crash, made a video to be aired in the House of Commons and had a run in with a knife wielding would-be kidnapper before being chased, in the dark, down a mountain by rabid dogs and into the face of oncoming vehicles—we were taken to Petra via his cousins wedding and were happily introduced as the guests of honour.

Text from Susie's brother

"Wow—busy day then. I went to work..."

Susie's Diary—July 2009

The thing we have found about the King's Highway here in Jordan is that the Kings can't have been fond of going around hills. No. It was absolutely straight up, straight down and be done with it. There is something quite terrifying about speeding to the bottom of a deep Wadi (dried river bed). Not the acceleration, more the horrible and exhausting knowledge that you will soon be pedalling all the way back up the other side."

The King's Highway

What a few days. First stop, the incredible desert city of Petra, sawn out of the rock face and then the vertiginous slog to the Christian town of Madaba. Aside from the dramatic inclines, the journey was also an eventful one. There was the 'bike meets gravel catapulting incident', the 'being pelted with fruit situation' and the numerous invitations to take tea from policemen, shop owners and petrol station attendants. Three things, though, stand out:

1) Being blinded with frequency by profuse perspiration
2) Being awed constantly by the castles littering the route, and
3) Being the catalyst for a 'personal moment' for a man near Kerak

The hills were steep and unrelenting. At times over 10 miles straight up. Delight to find that the highway goes right through Wadi Mujib, locally promoted as: 'Jordan's miniature Grand Canyon', was muted. It is fair to say that I had perspired a little on previous days riding. This, though, was the first time my usually sufficient eyebrows failed entirely to deter the salty sweat cascading into my sun burnt eyeballs.

"Jamie I'm sorry. I have to stop."

"Why? What? You ok?"

"Yeah—no problem—it's just that I've actually blinded myself with glow..."

Despite our odious personal conditions, we were buoyed by the constant distraction of the crumbling castles and the Highway's lavish history. Significant even in pre-historic times as a trade route between Heliopolis, the Sun City of ancient Egypt, and the Euphrates River, its northern reaches saw prodigious battles during the Holy Wars of the Middle Ages. Catholic crusaders blessed by the Pope rode thousands of miles from Western Europe to restore Christian control of Jerusalem. Thousands of kilometres from home, they built mighty citadels and fought fearsomely with the Arabic forces of Saladin.

I spent many a distracted mile imagining exactly how the crumbling ruins of Shobak would have looked at the time, with its imposing stone walls, secret passages, dank prison cells and menacing towers.

Mostly playing the role of a warrior princess.

With its many ups and exhilarating downs, it was a truly stunning piece of road. We camped on the edge of vast ravines, flew round chalky mountainsides and rolled into cities steeped in antiquity, only to be handed chilled glasses of orange juice and sweet cups of tea from generous strangers. The people here were even more gracious than we could have imagined, and our interactions with the local populace were marred only slightly by a more explicit vision of a dodgy man on a tractor than we had ever anticipated.

"Jamie.... is he? No. No. He can't be..."

"Can't be what? Can't b...Oh my God! Eurgh. Yep. Yep—I'd say he's pleased to see us!"

At the Northern End of the Highway we hit the town of Madaba, where the visa battle began.

15 July—Amman, Jordan

Day 62—4947.95km

When planning a trip of this nature there are two things that really destroy your budget: visas and vaccinations. I can't remember the upfront cost of the latter but I remember it being almost as painful as the actual jabs. You begin playing 'disease roulette'.

"So in comparison to say, getting eaten by a shark, how likely is it that I'll get Japanese encephalitis?"

Visas remained a heart stopping expenditure all the way round, in terms of money, but also in terms of time. The big visa issue hanging over us from the start was that of China. Unable to get a Chinese tourist visa more than three months in advance, we had no choice but to leave London praying we could pick one up on route. Our first attempt to procure one was in the Jordanian capital of Amman.

It was our first failure.

Sh*t!

Though we had begged, pleaded, cajoled, flattered and whined nothing could move the impassive staff of the Chinese Embassy who insisted, quite simply, that as we were not residents of Jordan, we could not get a visa there. Silently I began to fret. This wasn't the most assuring of revelations. We weren't residents of Jordan, but we weren't residents of any other countries we were riding through either.

The Prince of Jordan

In July 2009, Prince Hassan bin Talal of Jordan, a founding member of Desertec, announced the creation of the Dii, an industry initiative to develop a sustainable energy supply from the deserts of the Middle East and North Africa (MENA).

Made up of the business leaders: ABB, ABENGOA Solar, Cevital, DESERTEC Foundation, Deutsche Bank, E.ON, HSH Nordbank, MAN Solar Millennium, Munich Re, M+W Zander, RWE, SCHOTT Solar and Siemens, the Dii agreed a goal to use solar and wind energy to *"satisfy a substantial part of the energy needs of the MENA countries and meet as much as 15% of Europe's electricity demand by 2050."*[9]

It had been my hope that Jordan's links with Desertec would bring about some meetings or visits, but though I had sent emails a plenty and researched the area, I could find little to ride past and my calls went unanswered. It was with great relief then that when we got to the internet I read that Prince Hassan bin Talal had been doing decidedly better.

In Jordan, with no solar news of our own to write home about, I was glad to be able to profile his efforts.

---

[9] Desertec.org

Syria & Lebanon

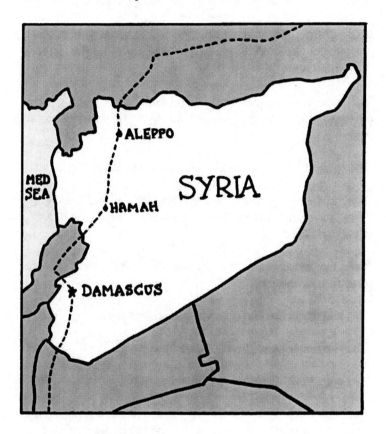

Estimated distance: 805km
Actual distance: 558km
Sunlight Hours per Year: Over 3000

Riding Tunes:

"You can't always get what you want
But if you try sometimes you might find
You get what you need"

You Can't Always Get What You Want, The Rolling Stones

17 July—Dar'a, on the Syrian Border

Day 64—5074.36km

Pushing thoughts of China aside, we headed for the Syrian border and promptly had more immediate immigration issues to contend with.

"So how was Israel Miss Wheeldon?" The heavily armed customs officer raised an eyebrow from behind the wooden desk.

"We haven't been to Israel." I answered, confused.

"Why not, don't you like Israel?" The official was having none of my 'innocent' charade and attempting to catch me in a game of extreme cunning....

"Erm, I've never really thought about it." Which didn't do much to allay my confusion.

"Did you like it the last time you went?" He persisted.

"But I've never been there!" I persisted back.

"Are you sure?" This was getting silly.

"Positive. Look. You can see here that we left Egypt and went straight to Jordan." I pointed at the stamps in my passport.

"Ah yes, why did you do that? Didn't you want to go back to Israel?"

Argh!!!

We spent an hour getting grilled at the Syrian border.

The army officers were either very suspicious, or very bored.

18 July—Damascus, Syria

The morning after, and persecuting only ourselves, we hit the road to Damascus. Uneventful other than an 8am puncture, directly preceding Jamie's 8am beer of frustration, we hit the city's outer limits and by mid-afternoon made our way through the melee of traffic to the bustling pandemonium of its centre. Within minutes we had done our utmost to hole ourselves into a suitably grimy hotel, get ripped off on the price tag and ensure that it sported not only a fine line in mosquitoes but the latest in strip lighting. A cold shower to scrape the dirt off later, we hit the bustling streets.

Time to explore.

Umayyad Mosque

As I looked around the Umayyad mosque, I was filled with a fleeting panic. Jamie was nowhere to be seen. A stranger from Iran had been peppering me with quick fire questions for the past 15 minutes, I interjected.

"I'm sorry, but I can't see my friend. I'd better look for him."

I wandered back through the middle of the mosque. The evening sunlight danced over the minarets and shone through the delicate gold lattice work. Rich mosaics blazed in its amber glow as the first shadows fell across the courtyard where children played hide and seek between the legs of their parents. A few people prayed but most were chatting idly and taking photographs. I scanned the scene.

Boll*cks.

The Umayyad mosque, home to the warrior Saladin's tomb, is the fourth holiest site in Islam. A reverently religious place, even women wearing headscarves are sent down an alley alongside to rent a burqa from the 'Special Clothing Room'. A few minutes earlier we had found this hilarious but now, like a tented, female 'Mr Anderson', everywhere I turned I saw women dressed exactly like me. Body after body covered head to toe in green, eyes and shoes the only things peeping out from cavernous material. I was suddenly lost in a sea of shrouds. With short brown hair and a Mediterranean complexion, Jamie was often mistaken for a local. I couldn't see him. There was no way he could pick me out of the crowd.

It took 35 minutes to locate one another in an area half the size of a football pitch and was with great relief that we dropped off the fabric prison and headed straight for the liberal dress codes of the city's Christian quarter.

Magical.

If you think of a Middle Eastern scene set in the narrow lanes around a Sultan's palace, you'd probably get an approximation of the winding passage ways, thick wooden doors, overhanging balconies and ornate shop fronts of Christian Damascus. The sweet scent of jasmine flecked the air and twinkling lights danced from multi-coloured windows, breaking through the shadows of its narrow stone walls. We meandered down the cobbled pathways and got lost in the back streets until there, in amongst churches and boutique urn shops we saw the unusual glimmer of glistening moisture on light brown glass. Could it be? Was it? A pub?

Heathens to the last, we made haste to one of its three tiny tables and, following a wee in what has to be the world's smallest toilet, got merrily drunk while the locals chatted up the buxom Russian barmaid. Weeks on the road and metabolisms hitting turbo meant it wasn't long before we were giggling and snorting like school children, uncharitably discussing the unlikely physics of her ample frame fitting into the meagre facilities. We could afford the hangover though. We weren't going anywhere fast.

Visa issue two was underway.

Iranian visa fears

Before we had left the UK, Iain had set about the long drawn out process of getting us Iranian visas. It was never going to be simple but none of us thought to question quite how hampered our plans might become if the country entered the throes of a revolution; which it verily attempted. In the intervening months an election had been called that the Iranian president was accused of rigging. Riots ensued and were blamed immediately on the interference of the UK. Which I thought was a little farfetched until someone pointed out that it was exactly such interference that led to the UK/CIA backed deposition of the democratically elected President in 1953.

Ok. Fair enough then.

In any case, British citizens were no longer being issued visas and the Foreign Office was warning against all travel to Iran. Iran, the country which sat squarely between us and Central Asia. Iran, the only way that we could get to Central Asia without a navigation of the Caspian sea... Or a short hop through Iraq.

And so it was that, with me chastely draped in a bed sheet from the hotel[10], we hot footed across town to chance our arms at getting our passports stamped. On the dot of nine we lined up with a handful of others in the cool embassy waiting room, trying to look inconspicuous and praying that the Iranian news channel took a temporary reprieve from its vitriolic assault on the morally pugnacious British. As our turn was called, I pushed our passports across the counter and accompanied them with my most beatific smile.

"Suse, stop trying to channel the Virgin Mary. It makes you look demented."

---

[10] Women are legally obliged to cover their hair in Iran and it is a prerequisite of entrance to the Iranian Embassy. At least if you want to leave in possession of an Iranian visa.

It worked though. Two days later we stood in the self-same spot, amazed, shocked and blinking at our newly accredited documentation. That couldn't be right, could it? There were restrictions on all entry of UK residents. Everyone else we'd heard of was refused admission. Surely it had been too easy?

Our joy was short-lived.

It had.

Incredibly we did have visas, but our visas were for the shortest entry time possible, 15 days. Iran is a large country with two mountain ranges to cross between its Turkish and Turkmen borders. I had liberally applied thumb to map and scheduled us 24 days to get from one side to the other, requiring a 30 day permit. With nine less, even my usually flagrant optimism was wavering. It was a long, long way to go under our own steam.

With no other option, we pushed off early the next morning. Having planned to traverse Syria to the East across the desert past Palmyra, there had been a swift reroute north. One as the former led directly to the disputed Kurdish region along the Iraqi border and two as we could now afford to.

Spot the mistake:

16 July, 17 July, 18 July, 19 July, 20 July, 30 July, 31 July

Yes, whilst route planning, I had cleverly failed to count from 20 to 30 and completely missed 10 days from the schedule. This and our now shortened time in Iran meant that we had time for a side trip.

The magnificent city of Baalbek lay 100 miles away in the North of the Lebanon.

'Baalbek' translates as 'Sun City'.

On a solar mission, it seemed reason enough for a detour.

Lebanon

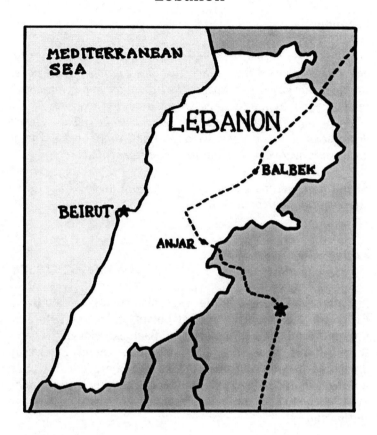

Estimated distance: 0km
Actual distance: 91km
Sunlight Hours per Year: Over 3000

Riding Tunes:

"With a few red lights and a few old beds
We made a place to sweat
No matter what we get out of this
I know, I know we'll never forget"

Smoke on the Water, Deep Purple

Jamie's Diary—August 2009

Lebanon

"The Lebanese border was typical. Like many other borders we had to run around from office to office getting bits of paper stamped, pay exit fees, return to the first office to get stamps for the bikes only to find there were no stamps for bikes etc. All the while re-joining queues that turn out not to be queues, clearly demonstrated by someone walking in and simply passing their documents over your shoulder to the official behind the desk.

Saying that, getting out of Syria definitely involved less of an interrogation than getting in."

22 July—Baalbek, Lebanon
Day 69—5289.91km

Just after the Lebanese border we took a swift diversion to eat ice cream and climb the ancient pillars of Anjar[11], before an uneventful ride to the Roman city of Baalbek. Though the gates were almost shut, the guys manning them generously presented us with half price tickets and offered to stay open until we wanted to leave. We bumbled around in the oncoming dusk, cocooned by the balmy evening and rusty rays of the setting sun.

Once again we had the place almost to ourselves, astounding not least because it is renowned as one of the most impressive sites in the Middle East.

"The sheer scale of the temples, columns, and altars is staggering. The fact that the small temple is still standing after two earthquakes which destroyed practically everything else indicates the magnitude of the architecture.[12]"

---

[11] The only Umayyad Islamic site in the Middle East
[12] Jamie's Diary

It was magnificent and we clambered all over it, chatting to the ten other people ambling around and wishing we had more than 24 hours to explore this incredible country. We could see Beirut, Byblos, Tyre... really though were just gutted we couldn't stay another week, when Baalbek was, randomly, to be the strangest site I could imagine for a Deep Purple concert.

Susie's Diary—July 2009

"Motorcyclists.

I don't know.

I was just getting settled into a happy routine. Fearless and unconcerned by the burning heat and traffic passing by at break neck speeds and what happens? Not one—but two—guys reach over and feel my bum as I ride past.

On the same morning!

The first, at least, was cheeky. A Bikers' Convention had escorted us as we cycled away from the Lebanese border. And—despite the death defying manoeuvres it took for them to joke around with us—the mood was light hearted and entertaining. When one reached across to slap me on the ar*e it was (largely) in jest and met with an amused reprimand.

The second guy was just a dodgy bloke who went for a full-on fondle.

Nice.

After the grope we found him parked further up the road, so stopped and confronted him. He stared back in incomprehension, no idea why his action was inappropriate.

Though I am ashamed at the lack of composure, it is fair to say he had a rather expressive introduction to the English language!"

North West Syria

From Baalbek it was back to Syria and Crac de Chevalier, the largest of the crusader castles dotting the region. The landscape was getting greener, the days cooler, and we had hit the tourist route. The holiday feel, last felt careering past the chateaus of Southern France, fleetingly returned. We ate by the huge waterwheels of Hama, gazed out from lofty battlements and caused mayhem getting spare parts in Aleppo, accompanied by most of the city's excitable residents[13].

Gliding past the vast cornfields of the Syrian plains we looked up to find Esse, our first long distance cyclist, pedalling the other way. Equipped with a rickety bike and laden with nothing like the latest in modern technology, he had simply gathered everything he could and set out on an adventure. Starting in Russia, he had headed south without a plan and followed the winds wherever they roamed. An incredible guy, he had not only upped sticks on a miniscule budget, but was also learning Arabic along the way. Staying with local families and taking the time to learn from them, he had already had a wealth of unbelievable experiences.

As we got back on our bikes I mused over the many different ways to travel. Some cycle for a record, some for a cause and others for the simple joy of exploring whatever the world has to offer.

Esse had an incredible lust for life and a desire to see as much of it as possible. I sometimes wonder where his journey took him.

---

[13] As I write this conflict has broken out across this whole region. The people here were kind, considerate and welcoming. I hope the fighting is short-lived and does not claim as many lives as feared.

Turkey

Estimated distance: 386km
Actual distance: 925km
Sunlight Hours per Year (in East): Over 3000

Riding Tunes:

"I got nine lives
Cat's eyes
Usin' every one of them and running wild"

Back in Black, AC/DC

26 July—Gaziantep, Turkey

Temporary reprieve from the Middle East could not have come sooner. Though the people had been generous, the scenery stunning and the sites more incredible than we ever imagined, I cannot tell you how good it was to see free flowing hair and women in strappy tops. Worryingly, I had started to notice a plunging neckline even before Jamie.

I hadn't expected to feel so relieved, nor realise how the attention had been weighing me down. To go into a room and have every eye watching you, watch you eat, watch you talk, watch you drink, watch you chattering. Watch you get up to the loo, blow your nose, put repellent on, laugh at a joke or to write in your diary. Not only was I looking more demure, but I had also begun to act more demurely. I would wrap my sarong around my shoulders, let Jamie talk with passers-by and be quieter than my usual self.

In most of the roadside cafes we entered, I was the only woman. On most of the streets we walked down, I was the only woman and in most of the encounters we had, I was the only woman.

In Turkey we were immediately surrounded by the usual babble of an excitable throng. But this time the throng was not just a throng of men. It was a throng of all types of people, men and women, young and old. I laughed and joked and did not flinch when they threw their arms around us. I felt new again, free again. And the joy of this intense relief was compounded by another overwhelming sight.

Grass!

This was the first time since the 28th May that we had stood on rich, succulent, green, green grass.

Which felt even better.

We drank beer, smoked hookahs and chatted to travellers and locals alike. We were welcomed at every turn with offers of tea, and had a once in a lifetime opportunity to eat as much baklava as physically possible without piling on the pounds. For those first few days in Turkey, it couldn't have been any better.

Susie's Diary—July 2009

Buoyed by the fantastic conditions—we set out on a 185km day to Mardin. Man alive! I think I have bruised my internal organs.

First things first though: my inaugural flat.

Jamie has not been blessed when it comes to punctures. At one point they were a daily occurrence. Having instigated the 'fix your own puncture rule' early on*, he would toil away unaided in the midday heat, while I did a spot of sunbathing.

In nearly 6000km I had escaped unscathed.

Until now.

It was the moment of truth. Had my long days in a shed with bearded cyclists and resultant bike maintenance NVQ equipped me for the situation...?

17.5 seconds later a gang of boys arrived and leapt to my assistance. I had just had chance to deftly remove the inner tube when it was forcibly taken from me, patched up with a flourish and my puncture repaired within moments. I was then kissed on the hand in a gallant manner and presented with my fully functioning bicycle.

Jamie looked on shaking his head.

"Jammy g*t!"

*The first time I tried to help him I nearly poked his eye out with a spoke. He never let me help again.

Foreign and Commonwealth Office—Travel Advice

Overwhelmed by the continued chivalry and kindness, things were, however, about to get a lot harder.

"We advise against all but essential travel in the provinces of Hakkari, Sirnak, Siirt and Tunceli and visitors should remain vigilant when travelling in other provinces in south eastern Turkey. Terrorist attacks are regularly carried out against the security forces in the south east of the country by the separatist Kurdistan Workers Party (PKK)."

Eastern Turkey

Having assessed the safety of our route using the fool proof checklist of 'anything other than Afghanistan, Iraq or the Pakistani border', finding that our path through the disputed Kurdish region of eastern Turkey had an FCO travel warning was unexpected.

In 1978 the Kurdistan Workers' Party was created to campaign for Kurdish rights. Following the incarceration and maltreatment of a number of its members and their suicide by hunger strike and self-immolation, the party turned to armed insurgency. Despite periods of ceasefire, attacks on the Turkish state continued throughout its Eastern provinces and, in the autumn of 2008, a violent clash saw around 40 deaths. Arriving during an armistice and taking all FCO warnings with a pinch of salt ('I'm sure they veer on the side of caution') we were nevertheless totally unprepared for what the Kurdish territorial dispute would mean for the average touring cyclist.

Pain!

Whilst the West of Turkey is the epitome of a quickly modernising nation, the East remains firmly in the developing bracket. And what they appeared to be developing specifically in the summer of 2009, was the road surfaces. Now, my checkered career has failed to include the practice of road building, but one thing became immediately apparent, even to a layman: if you want to build a road, somewhere where there is one, you have to take up the old one first.

Which, in the interim results—rather distressingly—in no road at all.

At one point my back wheel literally fell off my bike, and I am pretty sure there was internal bleeding. I don't get angry. It is well known. Only my mother can induce in me a sea of insurmountable rage. My mother, that is, and cycling down the roads of eastern Turkey.

"F***ing b*tch a*smoth*rf***er!"

At least the torrent of expletives kept Jamie entertained.

To compound the issue the east of Turkey is also pretty hilly; beautiful, but hilly. We clambered up from the Syrian plains and battled our way over vast ravines, sweat pouring from us, rabid dogs in persistent chase. We waved at the bemused road workers, drank endless cups of generously proffered sweet tea and eventually made it to Hasankeyf, a Troglodyte village built deep into the cliff face.

Heading north to the 'Cold City' of Tatvan the next morning, we were excited by the thought that it might provide respite, not only from the steep inclines, but also the long, hot, humid nights spent melting in claustrophobic tents. So much more though, we were excited about the town we would pass along the way.

BATMAN!

One mile before Batman, Turkey

Three elderly men sat on broken stools at the front of a petrol station. They were clearly settling in for a day of idle banter and watching the world pass by from their front seats at the transport hub. It was too much delight that entertainment rolled up in the unexpected guise of two dirty looking British cyclists, heaving their worldly belonging. The usual greetings and tea abounded. Maps were studied, destinations outlined and biscuits were eaten.

Standard.

These interactions happened all day every day and left us with not only full hearts, but as wired as ninjas on acid from the sheer volume of sugar rammed into the tiny tea glasses. Never before, though, had we added into the familiar routine, removing paper, scissors and pencils from our bags and creating a 'POW' sign complete with jagged comic book edges.

No matter how hard we tried to mime a fight between the Masked Crusader and the Forces of Evil, it was to no avail. The old gents, delighted by our enthusiastic exertions, had clearly never watched the 1960's classic cult TV show and were even more baffled when we had finished than before we began.

I doubt that a mile later by the Batman road sign, watching me punch Jamie in the face with the paper 'POW' attached to my fist, would have done much to help matters.

2 August—Tatvan, Turkey
Day 80—6376.83km

By the cool of Lake Van we took a day to check emails and sort out plans. While we had been staking out the Iranian and Chinese embassies we had also been working with a Central Asian tour agent to arrange our entry to the region. Now within spitting distance of Iran, we finally received the dates of our Turkmen visas and found that we could not enter the country until the 11th August. If we did, we would reach Turkmenistan too early to cross the border. With time to kill, we squandered hours in local cafes piling up on energy reserves and playing ever more jeopardous games of Scrabble on Jamie's phone.

"Jamie, you cannot have Za!"

"Yes I can. It's a word. It means pizza."

"If you went into a restaurant and asked for Za they'd spit on it."

"It's still a word."

"I cannot believe you actually learnt all the two letter words in the alphabet. How bored were you!"

Lured in by the neon lights of one of the many establishments, we looked through the menu as we continued our bickering. Experience had taught us that the advertised fare should always be taken with a pinch of salt, but we gave it a go anyway.

"Omelette please."

"Meat kebab madam?"

"Burger?"

"Chicken kebab madam?"

"Rice?"

"Tea and kebab madam?"

"I'll have a kebab, thanks."

"Yeah, me too."

We had a week to cover 150 miles, not much more than we had done in a single Libyan day. And so it transpired that, only 30 miles down the road to Muridaye the next morning, we saw the waterfalls, shoved the brakes on and without a second thought, slung off our panniers and pitched our tents. We bought a beer and started a poker game with the guys who ran a restaurant nearby, luxuriating under sunny skies and kicking back in the peaceful idyll.

It was wonderful to have the time to stop, to chat and to get beaten at cards. One guy styled himself as a magician, another brought out a puppy and they all laughed heartily at Jamie when he did the washing up, highly entertained that he would undertake a woman's task. We wandered to a nearby rope bridge and explored the area, passing the time of day with the local families that milled around the river banks and plunging into its chilly waters. We slept well in our tents in the cooler air and brewed up some coffee for a lazy breakfast.

The rest did not last long though. Having failed to remember to get out any money, we had only the emergency $100 my brother had hidden in a torch for us. We needed to get to town where we could get some supplies and begin to make plans for our Iranian crossing. There were other reasons to push on to the border, though. One, as we might have time to climb Mt Arafat[14] and two, as some Polish cyclists had arrived and passed on a warning.

While we had spent one of the most relaxing nights imaginable, they had been terrifyingly awakened by a man wielding a Kalashnikov.

A swift pedal through the area seemed prudent.

7 August—Dogubuyazit, Turkish border
Day 85—6648.23km

Though several people in the town were climbing Mt Arafat, it was not as simple as it seemed, requiring guides, permits and a desire to tramp for miles in particularly inclement weather. We were about to start a 15 day speed race across a dry nation and had just arrived in a place with alcohol. It is safe to say that not only did we fail to climb to the mountain's summit, but that

---

[14] Mt Arafat is the supposed resting spot for Noah's ark

little else was achieved during our stay in town. Aside from a quick visit to the stunning Ismak Pasa Palace, we hung out in the youth hostel, ate Za[15] and reviewed our depleted team dynamics.

Generally good but intermittently strained.

"Jamie, if you tell me one more time on the up hills that you aren't tired, haven't broken a sweat and were so bored you nearly fell asleep, I am going to physically bludgeon you to death with one of my spokes."

"I don't do that!"

"You did it yesterday."

"Ok. I did do it yesterday."

"And the day before that."

"Ok and maybe the day before that."

"And the day before the day before that."

"Ok, ok—but if you say you are going to stop at a shop but then stop AFTER the shop one more time...!"

---

[15] Seriously, come on... who says Za?!

Iran

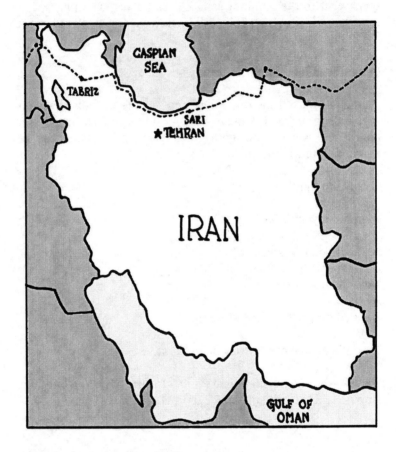

Estimated Distance: 1609km
Actual Distance: 1821km
Hours of Sunlight per Year: Over 3000

Riding Tunes:

"I read the news today, oh boy"

A Day in the Life, The Beatles

Iranian border

We pedalled up to the border with trepidation. Though our Iranian visas were safely nestled in our passports, we were still unsure if we would be allowed in and—following the arithmetically mind-bending complication of changing one foreign currency for another—made our way cautiously through the high metal gates to the immigration hall. Soon enough we were in the interview room, me now wearing a shawl to cover my hair up and a knee length overcoat bought in a Damascene market.

"Where are you going? Why are you here? What's in your bags?"

The minutes ticked by. In the intervening days we had spoken to everyone we could think of. All the travel advice had told us not to go to the country, a cyclist writing for the Telegraph had turned back, the news told of rioting and arrests, my friend with contacts at the Foreign Office told us to give the country a swerve...

"Welcome to Iran!"

We were in.

Just over the border we pulled into a roadside cafe. The TV was on in the background with a turban headed Mullah pontificating wildly.

"Look at that man. He's a terrorist. Ha ha—and he wants to kill you!"

They laughed.

We laughed.

Just a tad more nervously.

Cycling in Iran—Scene One

Luckily, it soon became abundantly clear that the only thing the Iranian people were going to kill us with, was kindness.

Picture the scene:

(Steep hillside descent. Two cyclists. One male. One female. Dirty, shell-shocked, gleaming from the humidity. The female is wearing three layers and a headscarf. A constant flow of traffic speeds past them at treacherous speeds. In nearly all cars the occupants are leaning out of windows, waving, holding up their mobiles, videoing and proffering sweets.)

CYCLIST ONE: Salaam!

(Cyclist one, waves, wobbles, clings to bike. Pothole.)

PASSENGER ONE: Salaam! Where are you from?

CYCLIST TWO: Hi, from England!

(Cyclist two, uses one hand to desperately grab solar charging phone as it escapes from its home on the bicycle and bounces over the handlebars. Pothole.)

PASSENGER TWO: Hello! Hello! Where are you going?

(Passenger Two scrambles precariously over Passenger One to get a better view.)

CYCLISTS: Erhgmpf!

(Both cyclists, wave and swerve into the road to avoid the gravel. Pothole.)

CYCLIST ONE: Argh!

(Both cyclists, swerve back to the gravel so they are not mangled by a passing lorry. Pothole.)

CYCLIST TWO: Mpfh. Turkmenistan!

(Both cyclists pull back out into traffic to get round a slow moving truck. The car drives on, passengers waving through the rear window. The next, previously behind the first vehicle now pulls alongside. Pothole.)

NEW PASSENGER: Salaam!

(The new car swerves erratically into the cyclists while a passenger pushes cake into the hand of Cyclist Two. Pothole.)

CYCLIST TWO: Argh! Umpf.

(Cyclist Two tries to take the cake but—due to the speed of the descent and the unfortunate lack of tarmac on the road surface—cannot manage whilst holding on to the bike with only one hand.)

PASSENGER FOUR: Can we film you?

(The car is inches away from the Cyclists. A sheer cliff face drops off to the other side.)

CYCLIST ONE: Of cOursE yoU cAn, nO problem.

(Pothole.)

And so it continued.

For days.

11 August—Evowgli, Iran
Day 89—6826.31km

We were like travelling royalty—pedalling celebrities. Everyone wanted their pictures taken with us, or for us to star in their family videos. We drew crowds and cars and bus loads. I'm pretty sure I signed an autograph. It was even more laughable in light of the reservations we'd had about even entering the country. From the very first night we had been side-swiped by the sheer exuberance of the Persian hospitality.

On our revised killer schedule it was well after dusk had fallen by the time we drew into the town of Evowgli. Hoping to find somewhere to lay our heads, it was dispiriting to ride up the only street of shuttered garages, closed tyre shops and barricaded hardware stores. With no hotels to be seen and nowhere to hide a tent on the barren horizon, we drew up alongside a restaurant to ask the owner if we could camp on the grass outside.

Much activity ensued. Maps were shown, tea was ordered and arms were thrown wide in greeting and exclamation. Minutes later, still unsure what was going on, I was plonked in the midst of the old men clucking around us and Jamie was marched off with the guy sporting the most prolific moustache, clearly demarcating him as the one in charge. More maps, more smiling, more tea until the pair returned; the arm of our beaming benefactor now slung around Jamie's shoulders.

We had a bed.

In the local hospital.

In the UK we have the Red Cross. In Iran it is the Red Crescent. The generous staff had an empty ward that night and had immediately offered it to the roving travellers passing through town. Minutes later we were being overwhelmed with soap, towels and ever more eye watering cups of sugar, topped up with a hint of tea. The medics, playing basketball rather than stemming the tide of medical emergency, were delighted by the distraction of two dirty English vagabonds. They bombarded us with questions and tested out our kit until one accidently caught me without my headscarf on and they took refuge in the kitchen, worried my honour had been mortally offended. It hadn't, of course, but after such a long adrenaline-fuelled day we were actually relieved to get some rest and—after the obligatory round of photographing ourselves pretending to be mental patients—fell fast into a deep and contented slumber.

Much to our horror, the next morning we caused a huge offense; something we would do time and time again throughout Iran. With only a 15 day allowance on our visa, we had calculated only one afternoon of respite before we hit Turkmenistan. With no contingency time and no time to stop, our distances needed to teeter around the 100 mile mark nearly every day we spent in the country. Without even thinking we had dressed, crammed biscuits into our mouths and packed our bags in order to leave.

"You're not staying for breakfast? But our friend has gone to buy supplies."

The medic was insistent.

"I am so sorry. But we have to go. We need to cycle a long way before nightfall."

"But he'll be back soon."

"I know, but it is 8am already and we really have no time."

"Then we will drive you some of the way."

"Oh, but we have to cycle."

"Really? Why?"

It is hard to explain the stubborn idiocy of the British cyclist on a long distance challenge, and our attempts to do so failed miserably. Why indeed would we cycle when we were pushed for time and someone was willing to feed us a hearty feast then give us a lift? Saddened to upset to our new found friends, we pedalled sheepishly into the morning mist and set our course for the Caspian Sea. The Caspian coast would be flat, flanked by settlements and allow us to ride alongside water before we turned to Central Asia.

It was August, so every holiday maker in the country had the same idea.

The Caspian coast, Iran

"Suse, are they refugees?"

"Um, I don't think so. It's just thousands of people camping…"

Our route via the Caspian was like riding through Glastonbury-on-Sea, if Glastonbury had no music, no alcohol and a significantly reduced number of people wearing hessian. There were tents as far as the eye could see. Big tents, little tents, green tents, blue tents, tents tented to other tents to create great big giant tents. And where there weren't tents, there were picnickers; picnickers on rugs with massive boxes of food, vast thermoses of tea, and arm-loads of fresh juicy fruit. We were beckoned over with sweets and chased by women brandishing cashew nuts. When dinner consists of tomatoes, pasta and garlic powder scraped from the bottom of your pannier, this is a welcome sight indeed and as time passed, we became more and more brazen in our actions.

"That family looks like they've got some curry going. Let's stop near them…"

And so our benefactors became more brazen too.

"Jamie—exactly how far has that car followed us now?"

"About 20 miles."

"Ok. That's ridiculous. We're gonna have to stop and have a picture taken."

"Seriously. Again?"

"Come on. We have to. They're never going to go away if we don't. Besides, they've been waving that melon at us for 15 miles and I'm hungry!"

Aside from the acrid smoke spewing from the trucks creaking in front and the acrid smell emanating from our armpits in the humidity, it made for an easy ride.

But I'll come back to that in a minute.

I got lunged!

Taking the affronts to my modesty up to five.

Outside the town of Anzali, a football hitting the roof of my tent preceded the alarm (danger of camping on the pitch) and we had made the most of the cool morning to race to Ramsar, boosted by flat terrain, tailwinds and decent roads. At our lunch stop the genial owner refused payment and only conceded our departure following photographs with every one of the staff and providing we took fruit, bread and doggy bags away with us. Though we had pedalled a fair distance already, we set out to attempt another 100km to the Port town of Chalus, riding high on the great conditions and bonhomie.

Boom.

Splat.

Argh.

Well, I assume that was what happened.

Jamie was behind me when his tyre decided to leave his wheel, resulting in his stopping rather abruptly and a short flight across the handlebars. Fortunately the accident had happened while he was chatting to the driver of one of the three motorbikes lining up to ask him questions who then sped to catch me and furiously waved me down. Jamie was duly checked for damage and driven off, offending tyre in hand, to find a replacement. I was left with the other two motorcyclists; one when we stopped at a house nearby and the owner asked the other to help make tea.

Unfortunately, by this stage I had let it slip that Jamie and I were not married. Furthermore, I made the fatal error of explaining that Jamie was not my boyfriend and that we slept in separate tents. Once alone, the remaining rescuer pointed at me, then himself, before closing his eyes and pursing his lips together in the international signal for 'give us a snog'.

I declined.

Sadly, as I did so, I also laughed[16]. The 'no' was abruptly ignored and he dived across grabbing for my chest as he did so. Fortuitously, I can spot a lunge from 100 paces and swiftly deflected the advance. Attempting my best line in outrage, I scalded the offending young man who promptly turned crimson, realised that groping lone women wasn't de rigueur in Europe and apologised profusely for a long period of time. Then, thankfully, everyone else came back.

Anyway, it was nothing too worrying but I mention it as it is the reason for the terrible profusion of protracted lies we subsequently told.

Following the lunge we decided it was safer to pretend that Jamie and I were happily married. Not only to avoid more incidents, but also to make life easier in casual conversation, the question of our betrothal being the second most common after the interrogation over what on earth we were doing.

"Are you married?"

"Yes—of course."

"Do you have children?"

---

[16] Come on, it was funny. A grown man was making kissing noises.

"Absolutely—but we don't like them so we've left them at home."

Which wouldn't have been a problem if I wasn't such an awful liar.

I detest lying, so am particularly bad at it. It isn't too terrible if it's a quick white lie, but when someone has invited you to their home and the deception gets worse and worse and worse... Jamie, who has no problem dallying with the truth, thought all his Christmases had arrived at once; singing, dancing and handing out dollar bills. After many months on the road, he had suddenly been presented with a new and highly entertaining way of amusing himself. We would be asked about love, family and plans for the future and he would fabricate ever more fanciful tales of our nuptials, all the while watching me squirm through mischievous eyes and a barely suppressed smile. The more I looked uncomfortable, the more he took to the task with relish, claiming we had been married by Elvis and making subtly sardonic comments about his 'angelic wife'.

Bas*ard.

23 August—Bojnurd, Iran
Day 101—8262.43km

None of the entertaining chance encounters were as awkward, though, as the night in Bojnurd.

On the road to Bojnurd we had been passed by a cycling road race, and one of the mechanics had spotted us and waved. As he returned home on his motorbike he passed us again but this time pulled his bike alongside us to talk to Jamie. There was nothing unusual in this. All day, every day, I would pedal along in peace and serenity as motorcyclists joined a queue to ask Jamie where he was from, where we were going and how long we had been together. Often they would ask to take pictures of us or invite us to stay. The conversations would consist of the repeated names of countries we were cycling through, and a protracted set of apologies for being unable to stop for a couple of days/weeks/months or years at the motorcyclist's home. For once, though, our kind companion lived in the town on route and as an added bonus had offered to give our bikes a once-over. It was fantastic good fortune and we were delighted to accept.

In the Iranian style, his parents' house had no furniture to speak of. Following showers, bike checks and a speed tour of the local town, we returned to find a feast had been laid out for us on the carpet and a number of family members invited round for the occasion. One of these was a cousin of the family who had excellent English and was brought in as an interpreter. She was a bright, attractive and eloquent woman and, unfortunately for me, a wistful romantic.

"But I want to know exactly HOW you fell in love. Where were you? When was it?"

I choked on a grape to get out of answering. I couldn't look at Jamie, knowing that one of us would cave and descend into a spluttering mess of laughter. The cousin would not give up.

"What did it FEEL like to fall in love?"

Oh God.

Later, with cushions now spread across the floor to sleep on, I curled up, relieved that the ordeal was over. It had been a hilarious evening, but at the same time it felt ungracious to have lied time and time again to a family who had offered us nothing but well intentioned kindness. I hated to think how offended they would have been to find that we had shared a room, unmarried, in their home and that our wedding had not been held in Vegas, just before our trip, on the 1st April...

To make things a spot more interesting, half way across the country, we hit Ramadan.

In previous days, the incredible hospitality of the Iranian people and the volume of campers meant that no matter where we stopped, within moments we would be surrounded by men and women offering hot drinks and honey-covered flat bread. During Ramadan, Muslims fast between the hours of sunrise and sunset, a particularly long period of time in the height of the summer and, I would imagine, potentially life threatening for anyone living in Northern Finland. As travellers we were exempt from the restriction but could not help but wonder what other obstacles were going to get thrown in our path. Being unable to buy food and water was certainly a new one.

Though Iran was a true education in unparalleled chivalry— "My dear Lady, it was my privilege to be of service", is not something you often hear getting directions around the M25— crossing the country in 15 days wearing hijab during the summer, Ramadan, and the beginnings of a revolution was not, perhaps, ideal.

Thwarted again in our solar efforts, Iran was also proving problematic for our kit.

Email to G24 Innovations—a year earlier

Dear Sir / Madam,

Next spring a friend and I will be cycling round the world, using solar power to charge our equipment. The idea of the trip is to promote renewable energy and climate change solutions in the run up to the UN Copenhagen Climate Change Conference.

I have been given your company details from a professor at the EPFL[17] and am writing in the hope you can help us find the best panels and batteries for an expedition of this nature.

If you are able to contact me or to provide the details of whom I can discuss this with further I would be extremely grateful.

---

[17] Ecole Polytechnique Federale de Lausanne

A year earlier, having found that nano-solar technology was pioneered in my friend's laboratory in Switzerland, we set her on a mission to track down a contact.

"Suse—a guy with a massive beard said you should call the lab sponsors, G24i. They've got an office in Wales."

I got to work.

After a couple of emails to the headquarters I had located their marketing manager, who not only loved the idea of our trip, but helpfully lived round the corner from my house. A quick chat in the pub later and G24 Innovations were behind the project. But they hadn't just come up trumps with the panels; they had agreed to custom build them into our bike panniers.

Nano-solar technology

Co-invented by Professor Michael Graetzel at the Swiss Federal Institute of Technology in Lausanne, dye-sensitised solar cells use two layers of silicon, creating a highly porous layer of titanium dioxide nanoparticles. These cells are more flexible than traditional solar cells, work in higher temperatures than traditional cells and are able to operate in low-light conditions. The most important benefit, though, is the increased size of the interface between the layers of silicon which allows more light, or photons, to pass through, resulting in a greater potential to create energy. Because of its porous nature, the surface area of the cell covered by the dye is a thousand times greater than the apparent area of the cell.

Or... nano-solar cells create a lot of energy (check), work in high temperatures such as the Sahara (check), in low-light conditions such as a rainy English morning (check) and are flexible enough to be sewn onto a bike pannier (check). They were perfect for our expedition.

Weeks later I found myself pedalling to the G24i Head Office in Wales where a lovely lady had painstakingly hand-sewn the panels into the saddle bags we had posted down to her a week earlier. With great fanfare and much excitement she unveiled her handiwork.

"Oh wow. They are amazing. Seriously... I didn't even know you could get floral multi-coloured panels?"

We had power, but the guys were going to kill me!

With our own solar power source, though, we were now able to use the Nokia GPS kit to get tracked, in real time, anywhere in the word.

Nano-solar communications in Iran

Anywhere, that is, that you can use an international phone.

With Iran on the brink of revolution, our international phone reception, tracking and all communications were blocked as soon as we crossed the border.

After two weeks off-line in a country we had been advised not to enter, understandably there was a lot of relief when we finally popped up in Turkmenistan.

Turkmenistan

Estimated Distance: 724km
Actual Distance: 631km
Sunlight Hours per Year: Over 3000

Riding Tunes:

"Painting every scene, a Super 8 with no rewind
You glide along in your moving dream
The lines of life around your eyes are gouged with everything you find
When you find that nothing is what it seems"

The Ride, Alien Milk

25 August—Turkmenistan

I flung my headscarf off. Man did it feel liberating, a tiny piece of cloth the physical embodiment of repression. Already excited about investigating life in the ex-Soviet Central Asian republics, I was ecstatic to find Turkmenistan to be an eclectic mix of Asia, Europe and the Middle East. Jamie was even more delighted. Proximity to Iran meant that half the women dressed demurely in long skirts and brightly coloured headscarves, but as an ex-Soviet state, the other half were peroxide blonde, long-legged and sporting miniskirts. Scraping his jaw off the hotel floor, I turned to the tour guide.

"He's been on the road a while..."

As in Libya, Turkmenistan had a very restrictive visa policy; so if you were in the country longer than the five day transit period, you were required to have a guide. Finding no way round this rule we had contacted a local travel agency who had, quite astoundingly, found us a cycling one. As an added bonus, while we stopped to sort out some embassy issues, they had holed us up in a fancy hotel in town.

And what a crazy town.

Turkmenistan's capital, Ashgabat, is difficult to describe. Row upon row of opulent marble buildings line the pristine streets, manicured lawns are sheared by teams of attendants and pathways are cleared by a synchronised team of elderly women wielding wide-brimmed wicker broomsticks. Everywhere you turn there are vast golden doorways ripped straight from the pages of fairy tales, new-age sculptures which flash like space craft and monuments to honour the ex-dictator, Turkmenbasy. Statues of Turkmenbasy, parks dedicated to Turkmenbasy, fountains built for Turkmenbasy and, in a particularly Disneyland-esque display, a giant replica of Turkmenbasy's 'Little Green Book'.

In 1991 the Soviet Union was disbanded and, with no choice for their people, entire nations were handed over to the Communist Party leaders who just happened to be there at the time. In Turkmenistan, Saparmurat Niyazov was not going to be one to let the opportunity pass. Taking to the challenge with aplomb, he renamed himself 'Turkmenbasy', or 'leader of the Turkmen people' and promptly became notorious for his extravagant demands and strident personality cult. In case anyone was in doubt about his position, in 1999 he declared himself the 'Ruler for Life'.

Sadly for Niyazov that was not to be much longer. In 2007 Gurbanguly Berdymukhammedov succeeded the 'Ruler' after his sudden death. Unfortunately, for his people, it did not end the repression. While many had hoped for reform they had received only the slightest relaxation to the dictatorial regime. Behind the marble façade and neon sculptures, crumbling apartment blocks wrenched straight from *Bladerunner* sheltered the majority of the population.

Uzbek Embassy, Turkmenistan

"You flirt with the security guard and I'll hide in this bush."

Jamie valiantly leapt into the foliage as he chucked me the hospital pass. Begrudgingly I began an exhausted attempt at come hither eyes which could be more accurately described as 'wood chip has catapulted into my cornea and I am desperately trying to remove it without using my hands'. In any case, the effort was wasted. The subject of my attentions was otherwise engaged in a dispute between a fearsome woman and an angry policeman, while all around us scraps of paper danced from hand to hand, being plucked, pulled, grabbed, dragged and variously manhandled by a braying crowd.

In Ashgabat we needed to pick up visas for Uzbekistan and Kyrgyzstan, and to make another attempt on the Chinese Embassy. We had scheduled in a day for the activity but events were hardly going to plan. Arriving at the Uzbek Embassy at

6am, we had hoped we would be the first in the queue so that we could swing round to the Kyrgyzstan consulate, hop past the Chinese one and be done by lunchtime. Easy.

To our distinct disappointment, as dawn broke, several people were already loitering outside the Uzbek consulate, chain smoking cigarettes and eyeing each other with ill-disguised suspicion. We joined the back of what could have been a queue and watched as slowly but surely more and more desperate souls arrived. Four hours and several two-letter word disputes[18] later, the shutter on the booth by the gate opened. A man yawned widely and in a monotone that couldn't even be bothered to care less, called out names from a list on his desk top. Other bits of paper were handed lethargically to the security guards and from them on to the gathering throng. This immediately sparked confusion, pandemonium and then an all-out brawl.

"What's the deal with the paper?" I turned to a dejected man standing beside me out of the fray.

"You need to get your name on the list so you're in the queue."

"Oh—I'm not in the queue? Which bit of paper do I need?"

"I don't know."

"How long have you been here?"

"Six days..."

Six days? Six days? With another time-limited stay of entry, we only had eight to cross the country. Overwhelmed and fearing for safety and sanity, we aborted day one; resolving to regroup and plan a proper assault on the situation after a bit more sleep.

---

[18] Li = Chinese unit of distance. That's a Chinese word!

Day two and the finest reconnaissance techniques the *A-team* could teach us were hitched up, rolled out and swung into action. I took on Face's role of wanton charmer and Jamie assumed the pivotal part of B.A. Baracus after he has drunk some 'milk' and gone to sleep somewhere off camera. It worked though; after only several hours of a nauseatingly shameless little-girl lost routine, the list was placed aside and we were shunted to the front of the queue.

Well, I was.

As the security guard motioned me through the gate, Jamie leapt up and scooted ran behind me, urged faster by the daggers plunging deep into his shoulders.

With 24 hours already lost, efforts at the Kyrgyz and Chinese embassies went even further from plan. The Chinese was once again an absolute defeat and at the Kyrgyz, despite Jamie's zealous courting of the busty administrator, the best she we could rustle up was a two-day 'Express Visa'.

We didn't have two days.

Leaving our passports at the embassy, we hatched a complicated rescue plan and made ready to roll out.

"I'm happy to come back and pick them up Suse..."

29 August—Ashgabat, Turkmenistan
Day 107—8469.12km

The temperature was easy on us as we left Ashgabat. Well rested, well fed and accompanied not only by our guide, Zhenya, but his brother Vitelli, we headed east. Spoilt by a wealth of professional support, we zigzagged carelessly down the smooth, wide thoroughfare, until, as if breaching the city's magical force-field, the road disappeared into an amalgamation of potholes to which we were more accustomed.

Nevertheless, in relation to Iran and the purgatory of eastern Turkey, the ride was plain sailing. We stopped for water by a local shrine, talked road racing with the guides and glided idly through the gentle miles. Gophers distracted us from our protesting back bones, and mountains cast elegant shadows deep across the valley floor. When Vitelli left us to turn back we pedalled through two more towns to a teashop where Zhenya joked merrily with passing motorists and shook hands with the friendly proprietor. We were showered with fizzy drinks and stuffed with stale biscuits. In his eagerness to please, the owner heartily enthused about our fitness and gave directions to another teashop ten miles further down the road. There was an hour left in the sun so we decided to push on before setting up camp.

35km, zero tea shops and a road vaguely held together by a smattering of tarmac later, we stopped to get directions from a man fixing his car in the absolute darkness. As we did so, a scourge of mosquitos descended upon us, viciously assaulting every single uncovered part of our bodies and having a go at the bits that weren't. Cars beeped in horror as they encountered us in the inky blackness, our bicycle lights pitifully inadequate for fending off the lorries thundering over the chasms at breakneck speed. The motorist explained that the next village was another 30km away, but that there was a truck stop 3km to the left, down a track barely visible in the vacuum of the night. We plunged down it, clattering through craters, chased by dogs and praying that he was right.

30 minutes later Zhenya added a 15th sugar to his first cup of tea. We had pitched our tents outside the strip lighting of a cafe nestled amongst the hubbub of the lorry park and gone in search of anything to shove down ourselves. In the beer bottle strewn pit stop, we made for a motley crew. I looked out of place, Jamie like a mosquito's pin cushion and Zhenya like a Thunderbird without its strings. I nipped to the loo and immediately regretted the decision. It was certainly not the first time I had recoiled in horror in a public toilet, but sense should have told me that a bathroom previously used by over 300 men on long distance truck journeys should be avoided.

Susie's Diary—August 2009

We awoke marginally refreshed and set out for what is probably the most disheartening cycling experience we have yet endured. Though we rose at six, by 11 we had only gone 30km. The potholes have taken their toll on Zhenya's bike, battering it mercilessly and eventually serrating his derailleur entirely off. An able mechanic, he deftly removed the trailing parts and put the chain back.

But though his bike can now move, we have a lot of miles left to pedal.

He is stuck in one gear...

30 August—Tejen, Turkmenistan

Day 108—8691.4km

The wind rose to a beating strength. We tried to use each other's slipstreams to maximise speed, but at best this was 15km per hour and, when I was in front, about 12. The rest of the day was a blur. We were all struggling against the gale force gusts, in pain from the constant jarring and battling exhaustion from the night in the truck stop. There were some small mercies—cycling behind a lorry for 10km, a teahouse with freshly baked pasties and the wind dying down in the evening—but largely it wasn't the best of fun. To top it all off, just when things were looking up, one of the spokes pulled through Jamie's wheel rim, cracking and buckling it.

Bugg*r! [19]

The next day was more of the same, only this time accompanied by puncture after puncture as we navigated the tarmac rodeo. In a bid to stop his wheel from splitting entirely we swapped my lighter panniers for Jamie's and our pace

---

[19] I can't resist, I've got to mention it. To add insult to injury, after all of that purgatory, Jamie woke up the next morning with a spider bite on his nether regions. Very, very funny.

dipped even lower. Jamie now spent the day battling headwinds up front while I toiled heavily in his wake. Zhenya sat in my slipstream; head down, shoulders sagging. There wasn't a hope that we could get Jamie a new wheel. With sports all but banned under Turkmenbasy's rule, there were no spare parts that could be found locally. Our only hope was that the rim would hold out to Uzbekistan where Iain, re-joining us for a few weeks, could bring out a new one.

But though Jamie's bike was in dire straits, Zhenya's was practically in pieces. He struggled on regardless, brow furrowed, teeth gritted in sheer determination. As a young man he and his brother had competed in cycling races across the Soviet Union and Iran, but for the last 20 he had done nothing more than pedal around town, do the shopping and occasionally break into a light sweat. He was astounding both Jamie and I with his tenacity but the constant repairs, embassy issues and reduced speed loomed large against the inky black mark stamped heavy in our passports.

Our exit date was in five days' time.

Susie's Diary—August 2009

It took every last ounce of strength we had to get to Mary in time for the overnight train back to Ashgabat. With minutes to spare we trailed Zhenya's sagging body through the station, palmed our bikes off on one of his friends and crumpled into our battered berths.

Eight hours later, blinking sleep from our reluctant eyes we were dodging the capital's rush hour traffic and hailing a cab for the Kyrgyz Embassy.

It was closed.

The Consul was out.

Are you kidding me?

2 September—Mary, Turkmenistan

Day 111—8841.05km

Luckily, our panic was only brief. Two hours later the Consul arrived and, newly stamped passports in hand, we were soon on the four-hour bus trip back to Mary. Back, that is, down the decimated road we had just spent three soul-destroying days cycling along. We passed the time playing more Scrabble and debating at exactly what point the driver would skid off the road and kill us with his erratic motoring skills.

It was after nine and pitch black when we finally arrived in Mary. We trudged across town to retrieve our bikes and once again found ourselves pedalling blind down potholed streets, winded time and time again as we plummeted into the chasms lurking in the tarmac. At 10.30pm we finally reached a hotel. Trying to maintain a semblance of normality, I then spent the next two hours hand-washing the sweat and mud from my greying T-shirts.

I nodded off sometime before 1am.

The alarm went at six.

At 7.30 Jamie and I were up, fed and ready to hit the road.

Zhenya was not.

His back brakes had gone the way of his chain. We watched the billowing winds increase as he worked on repairing his bike. Zhenya, exhausted, was clearly doing the best he could in the terrible conditions. But we hadn't foreseen the frustration a new and slower member of the team would bring.

East of Mary, Turkmenistan

For six weeks Jamie and I had cycled alone. I would be lying to say that this was always plain sailing but after so long on the road we had become ridiculously attuned to each other's wants, needs and every foible. Jamie wanted to stop and relieve his chaffing buttocks at least every two hours. I would miss all useful signs and stops if I was daydreaming about winning Strictly Come Dancing with a particularly dramatic Tango demonstration. Jamie put up with my reluctance to look after ourselves and our kit. I put up with him lecturing me on how we should look after ourselves and our kit.

With such a positive experience with Lamin and Mohammed in Libya, we had not considered how a new guide would affect the team dynamics.

Zhenya was a lovely guy. He was putting in everything he had, much of what he hadn't and whatever else he could gain from unfeasible amounts of glucose. He was generous, determined and tried his best to understand what we needed in order to help us.

He was annoying the bejezus out of us!

Thankfully, this side of Mary the roads improved and so did our spirits. We ate melon with old men, rode on donkeys by the roadside and were given Snickers bars by two matronly ladies flashing wicked smiles. We pedalled past camels, ran down sand dunes and filled our bellies with steaming bowls of spicy dumplings. The last morning, we rose at 5am from the porch outside the teahouse and packed up for our final 140km to the border.

Our permits were due to expire at 6pm that evening.

20km Zhenya got his first puncture.

28km Another spoke on Jamie's wheel broke

44km Jamie got his first puncture

71km We ran out of water...

Susie's Diary—August 2009

We have been cycling for quite a while now and make sure of certain things, mainly, where we can get food, water and shelter.

Though Zhenya was a great guy his English was very broken, meaning that he often had trouble understanding some of our questions. When we asked him to check where we could next get supplies he had insisted that there was a village ahead.

There wasn't.

Desert, Turkmenistan

Two hours later, parched from the relentless desert sun and fighting the urge to panic, we heaved our bikes down a sand engulfed pathway, calling out amongst the desolate concrete buildings. Nothing.

It was a ghost town, as if everyone had fled.

No children, no voices, no belongings in the deserted houses.

We pushed on to the railway tracks through the loaded silence; thirst pushing us to the edge of panic. Two men sat in the cool of the station starring at heavy ledgers under the lethargic thwack of a wobbling ceiling fan. What a posting. They looked up, shocked by the wild-eyed dirty entourage that had emerged before them, curious to find just where we had come from. Zhenya explained our predicament and they immediately set about retrieving water from the well. We collapsed under the shade of the platform, staring out at the rusty fenders and willing them speed in their return. We choked on the water when they did. Throwing it down our throats and splashing it on our faces. We had no time to rest though. There were still miles to go.

Filling our bottles we thanked the men profusely and got back on the burning road to Turkmenabat.

Susie's Diary—August 2009

"At 5pm, shattered and following a blistering race against time we arrived at the border.

Except we hadn't.

Zhenya had got the distances wrong again. This was a police checkpoint and the border another 26km.

Sh*t!

Our visa detailed this as our last day in the country. Frenzied, we started to hitch. Zhenya was totally confused. We tried again and again to explain that we had to leave before 6pm. He looked wounded. There was much high octane kerfuffle. Eventually a car was located that could just about fit Jamie, me and the bikes in. Waving a frantic goodbye I clambered with both panniers onto Jamie's lap as the car slammed towards the border.

We arrived 10 minutes late.

7 soldiers guarded the gate and the lights were off.

No amount of newly-applied mascara was going to get us across.

3 September—Farab, Turkmen border
Day 112—9099.87km

Having just missed the border crossing we resigned ourselves to a night as illegal aliens, settled in at a nearby teahouse and lamented the now futile decision to get into a car. Aside from the plague, Jamie's illness and my near kidnapping we had stayed true to a purist attempt to pedal every last mile of our expedition; no matter how cold, how hot, how tired. I had to admit, though, the high-speed border race had felt good.

Like an adrenaline fuelled escape attempt.

Which is probably why it then felt so bad when, through the low hum of voices on the neighbouring day beds, we noticed that Zhenya had arrived. The poor guy. Unwilling to leave us, he had resiliently given chase, his wheel totally buckled and his will destroyed.

A shell of a man, he sat heavily, smiled weakly and promptly necked a large mug of vodka.

Uzbekistan

Estimated Distance: 966km
Actual Distance: 1144km
Sunlight Hours per Year in Uzbekistan: 2000—3000

Riding Tunes:

"There's a blaze of light in every word
It doesn't matter which you heard"

Hallelujah, Leonard Cohen

4 September—Alat, Uzbekistan border

"Suse, why aren't you wearing your helmet?"

"Argh—it just feels so restrictive!"

"What do you mean? That's stupid."

"I don't care."

Protracted rigmarole and extreme flirtation later, the laughing border guards handed back our passports and we pushed our bikes away from the immigration hall at double speed. More concerned about the random electrical appliances in our luggage, the customs officers had let us out of Turkmenistan unimpeded by our expired visas and our hearts slowly dropped from our throats as we emerged into Uzbekistan.

We were free.

Free from races to borders, free from ridiculous headscarves, free from people telling us what to do and where to go.

After the harsh conditions of the windswept Turkmen desert, a feeling of calm slowly permeated through me; we sailed past well-kept gardens, ambled past whitewashed buildings and stopped under trees cunningly imitating laundry racks. Slowly the pressure I hadn't even noticed began to ebb away and with it every reserve of energy I had.

We had hurtled across two countries, fought for visas in numerous embassies and spent adrenaline fuelled mornings bricking ourselves at border crossings. We had dragged ourselves over mountain ranges, toiled through deserts and had our bikes give way underneath us with the constant jarring of pothole after pothole. We had been perilously close to death on numerous occasions, stuck twice in the desert without water and barraged with constant attention. In our brief moments of respite we had lain in our sweat soaked tents, desperately battling for sleep and being massacred by mosquitoes. A blanket of foggy tiredness settled between my temples. We would soon be in Bukhara, one of the most astounding cities in the world.

And we would have a day off.

February 2009—(six months earlier)

"Tell me again where you are going on your trip Susie?"

"France, Tunisia, Libya, Egypt, Jordan, Syria, Turkey, Iran, Turkmenistan, Uzbekistan..."

"Uzbekistan? When?"

"Um—not sure really. I guess sometime in August."

"Gosh. I'm going with a girlfriend of mine around then. Wouldn't it be marvellous if we were there at the same time!"

You know those people you meet in life you cannot help but fall in love with? Warm, funny and full of the haphazard kind of wonderfulness that makes you delight in the idiosyncrasies of humankind. Zani, my friend's mother, is one of those people.

4 September—Bukhara, Uzbekistan
Day 113—9195.44km

Jaws on our handlebars, we picked our way through the wide, majestic streets of Bukhara. As the call to prayer heralded the onset of dusk, lofty minarets cast long shadows across our path and blue mosaics glinted magenta on dome topped buildings. Donkeys pulled carts down sandy streets as stall holders called out proffering huge Soviet style fur hats and a riot of colourful fabrics. It was a sensory overload. I could not have imagined the magnificence of this city in the heart of the Silk Route, nor could I have imagined how overwhelming it would be to see

Zani after the nerve-wracking pandemonium of the previous weeks.

"Darling. Look at you. Goodness, I didn't realise you were cycling all the way from Turkmenistan today. You need to get some rest. We've found you a glorious hotel. Come along."

I nearly cried.

Running from Rome to Nary, the ancient capital of Japan, the Silk Route, or routes, are the most important trading conduit in human history. Well-trodden by the vast armies of Alexander the Great, these arteries linked Europe to Asia and, just before Christ made an appearance, were used to transport silk from the East for the togas of Roman senators. Historically the exquisite city of Bukhara lay in the midst of this pivotal caravan route and was fought over by countless tribes and empires, including my favourite warrior king—if you're allowed favourites—Genghis Khan. More recently though, trains, planes and container ships have made the transport of goods from China across the plains of Central Asia a bit on the protracted side. So Bukhara's astounding Mussulman buildings lie quiet but for the mutter of European tourists and hawkers selling old Russian army uniforms from the Bolshevik invasion.

Three hours after we arrived exhausted, sweaty and aching in muscles we didn't even know we had, we were sipping great mugs of beer and talking animated rubbish under the stars and minarets at Lab-i Hauz. Away from the busy streets and the bustle of the bazaars, Bukhara's central pond hummed with the cadence of musicians, the chatter of passing travellers and the soft murmurings of old men playing backgammon. We met other cyclists, heard news of our route, laughed late into the night and stumbled home through a doorway hidden between the trellises in a high stonewall.

Four days, five chocolate bars and another train ride later, we would be back.

Iain had returned.

5 September—Tashkent, Uzbekistan

Zani's was not the only friendly face we were due to see that week. Not only did Hans and Ayesha, friends of my parents, live in the Uzbek capital, Tashkent, but from Bukhara we would hop on the train to their house where Iain would once again be joining us.

Having wooed his fair lady back from the evil clutches of her upstairs neighbour, our erstwhile teammate had promptly begun work on meeting us along the Silk Route. His would be a 20 day sojourn back into the expedition before embarking on a new career in carbon trading and barricading the doorway to his girlfriend's apartment.

Iain's return sparked much excitement. One, as it would be phenomenally entertaining to have him back and two, as he was bearing our much needed supplies and spare parts. He arrived in the middle of the night and so our breakfast was full of overexcited chatter and followed by a gleeful plunder of his panniers. Jamie set about changing his wheel while Iain and I went for a walk.

"So come on then, what's the latest with you and Jamie?"

"Why are you asking that with a grin?"

"Oh come on, it's what everyone's asking!"

Getting to Han's and Ayesha's was a massive relief for another reason: The China Visa, our Bête Noire.

Chinese Embassy Visa Information

"(i) Please apply for a visa about one month in advance of your intended date of entry into China, and do not apply 3 months earlier than your intended date of entry into China. You should take upon yourself any consequences resulting from your failure to submit visa application at an appropriate time, which may

lead to either your already-issued visa becoming expired or it would be too late to get a visa before your planned departure date."

As we had not been able to apply for a permit before leaving the UK, or pick one up on route, we were now in a bit of a quandary. We had tried in England, Amman, Damascus, Turkmenistan and now in Uzbekistan but had failed in every effort get a Chinese visa.

Now two weeks from the border, we still had no way of getting in.

Sh*t!

China is a massive country that made up almost a quarter of our trip; to miss it out was simply not at option. Various plans were floated. Couriering our passports home was rejected due to the vagaries of the unfathomable Uzbek holiday system[20], as was a mission to Astana in Kazakhstan. We had begun to wonder if we'd have to go home to get them, when fate popped up to intervene. One of Zani's friends would be in Tashkent the following day before travelling back to England. My parents had scheduled a mid-trip visit and were coming to visit five days later. We could ask Timmy (knee now fixed and working as a stand-up comedian in London), to get our passports from the airport, take them to the Chinese Embassy, then—once stamped— pass them to my folks who could bring them back to Uzbekistan. My cousin, Amy, with whom my parents were staying, had an office next to the fire station in which Ian, boyfriend of Sophie, house mate of Timmy, was working. (Stay with me here.) The passports could get passed on down the chain and back to us.

It couldn't fail...

---

[20] There are often unexpected national days off in Uzbekistan on which absolutely nothing works for an unscheduled period of time for no apparent reason whatsoever.

Text from Sophie

"Hey Suse, I don't know how you get to be so lucky but your passports are now with your cousin to give to your parents. Oh—and a bra. (Why do I always have to buy you bras?!) Which has probably been worn by half of Southwark Fire station..."

Uzbek regulations

While the visa bedlam unfolded, we crossed our fingers and set about continuing the journey. Having taken the train from Bukhara to Tashkent we would be heading back there with Iain in order to cycle the section we'd missed. Though we knew that Uzbekistan was not the easiest country to travel in, we hadn't foreseen too many problems with the plan. Our host, Hans, did not feel the same.

"I've written a letter in Russian explaining your trip and got a declaration from the police that you've been staying with us, but I couldn't find anyone who would legally verify the photocopies of your passports."

"Is that bad?"

"Well, you see guys; it's illegal to travel without papers so the military might stop you... Are you sure you want to go back to Bukhara before you get your passports back?"

Hans had done everything he could think of to help us but to no avail. He was concerned. In Uzbekistan, until recently sanctioned for its oppressive regime, it is compulsory for foreigners to carry their passports at all times. Tourists can only stay in authorised hotels and international visitors are not even allowed to stay with friends without permission from the local authorities.

We were camping in fields, with no papers and a depreciating hope that my thinning T-shirt would get us past the abundance of armed check points lining the route. I made light of the situation but was secretly worried.

"Ah, I'm sure it will be fine. What's the worst that could happen?"

BBC Article—Uzbekistan 'prison torture' claim

"The Human Rights Watch report says that beatings with truncheons and bottles filled with water, electric shocks, sexual humiliation and threats of physical harm to relatives are the most common forms of torture. In the past there has also been evidence that at least one prisoner was boiled to death in an Uzbek jail."

11 September—Naiway, Uzbekistan
Day 120—9313.89km

"Ok, just keep cycling. Wave. Smile. Oh sh*t they are flagging us down. Suse, you go first. And smile more."

"I am smiling Iain!"

"Well, pull your top down a bit..."

We made it from Bukhara to Naiway with only a few heart-stopping incidents. Road blocks and an unprecedented puncture rate aside, it was an entertaining ride; Jamie and I giving Iain grief for the 'extra tyre' he had smuggled under his lycra shirt and Iain reaffirming his role as the outfit's chief flirt and luxury magnet. After a hilarious interlude when a young woman exclaimed that meeting him was the best thing that ever happened to her, our erstwhile co-rider campaigned strongly for wheedling our way into a hotel for the night and—as owner of the outfit's sole passport—was duly dispatched to do his worst. Paperless, Jamie and I waited outside the only establishment in town.

And waited.

And waited.

After about 20 minutes we locked the bikes up and went to investigate. Iain was in full swing. The 'occupation' question on the check-in form had been met with much enthusiasm and the assembled staff were being visually violated by our teammate rubbing his nipples and sashaying down the hallway. Iain, it seemed, had decided his life in finance was much less glamorous than insisting he was an international Speedo model. For reasons I still fail to fathom, the receptionist was besotted, and with some general doe-eyed flattery, Hans' letter and the passport photocopies, she finally consented to let us stay.

Several hours later, our departure from the establishment was with more than the usual vigour from our resident lothario. The stridently enamoured lady, convinced he was a better prospect than her current beau, had proposed marriage and was not willing to accept no as an answer. Iain made blistering pace down the road to Samarkand away from the smitten receptionist whose boyfriend was not only a particularly large man but 'probably in the Russian mafia'.

Iain's Diary—September 2009

Leaving town we were flanked by the first of many tens of miles of cotton fields. Uzbekistan is the world's second largest producer of cotton. Known as white gold in Soviet times, cotton has been a key factor in the demise of the Aral Sea. Uzbekistan is one of only two double landlocked countries in the world (the other is Liechtenstein—in case you ever get asked in a pub quiz) and much of the country is desert. Cotton needs large quantities of water to grow and during the 60's the Soviets started diverting the rivers that fed the Aral Sea into vast irrigation schemes.

Since those days, the Aral Sea has been shrinking and the scale of the environmental catastrophe has been growing. It is now less than 10% of its size in 1960 and is five times as salty, which has killed many of the fish. These fish were once caught by boats which are now rusting carcasses nearly 100 miles from the original shore line. To compound the issue, an island in the middle of the sea was used as a military testing site in Soviet times and stories now abound of dumped nuclear waste, weapons and rusting barrels of anthrax being revealed by the falling waters.

Once we heard all this we found there was a renewed spring in our step as we pedalled in the opposite direction.

Continuing east and fuelled by vast amounts of melon and meat of questionable source, we arrived in Samarkand, one of the oldest inhabited cities in the world and another favourite of the chaps at UNESCO. It has some incredible architecture and the Registan, a collection of three madrasahs (medieval Muslim clergy academies) are definite contenders for some of the most impressive buildings I've ever seen.

'Registan' apparently means 'sandy place' in Persian, which is understandable given it's a pretty dry country. However, legend has it that the sand was brought in to soak up the blood from the public executions that took place here up until the early 1900's.

It's the kind of place where you take notice of the signs telling you not to walk on the grass.

Samarkand is in a key strategic trade location and was a crucial point linking trade between China and the world to the West in days of old. Accordingly it has seen its fair share of dust-ups over time. Alexander the Great first muscled in on the act in about 330 BC and it seems various nasty blokes with sharp beards and even sharper swords loved nothing better than to roll up in town for a few rounds with who ever happened to be in the hot seat.

Samarkand is also firmly on the backpacker route and is one of the few places we have been through that has been. The hostel itself was historic in that certainly the bathrooms seemed to date from the early part of the city's 2,750 year old history. There were some colourful characters including a bearded communist Norwegian intent on committing the perfect fish related crime, a couple of lively Irish lads and two young ladies intent on each taking a bit of Jamie with them on their travels!

13 September—Samarkand, Uzbekistan
Day 122—9425.29km

In Samarkand, the bugs certainly took a bit of me.

I awoke the next morning mauled raw and brandishing a skin graft of evil welts. Iain, with whom I had shared a double bed, got three on his neck.

"Oh man they're so itchy."

"Itchy! Itchy! Don't talk to me about itchy. I've literally got 200 of the things."

"Yes, but I think they're itchier on a man..."

17 September—Tashkent, Uzbekistan
Day 126—9820.40km

Two days, one projectile vomit, three passport stand-offs and 25 punctures later, we made it back to Tashkent where Hans, Ayesha and their two beautiful daughters welcomed us with open arms. Everything was changing fast now, Iain had returned, we had been thrown into the heart of a welcoming family and, without chance to even think about it, I was now standing outside the airport, anxiously waiting for my parents to emerge.

Half way round the world, my folks had decided to come and visit and drop in on us intermittently as we headed east across the Tian Shan Mountains. Paranoid I would somehow miss them coming through the airport's only exit, I fidgeted from foot to foot and endlessly scanned the arrivals board.

"Susie! Susie! Over here! Over here!!!!" I needn't have worried.

My father watched in awe as my mother and I battled each other for air space.

"Oh Suse, you've lost so much weight. What are all those red bumps? How's Jamie? When did you get here? How are the bikes? Is Iain back? How's Ayesha? How's Hans? Where are you going next? When are you leaving? Do you know where we're going? Do you want your nails done?[21]"

That evening, sated after a massive feast, we snorted with laughter under the twinkling candlelight of Hans and Ayesha's garden. Iain recounted his run-in with a domineering Russian stewardess and my father poured over the maps to plan where our routes collided.

Suddenly amongst some of the people I loved the most, I was on top of the world. Because of them all, of course, but also because—finally—we had spanking new Chinese visas emblazoned across our passports.

The Chatkal Mountains

Loaded with UK sourced biscuits and lined with red wine, we sped off the next morning on a high-octane sugar rush to Kyrgyzstan.

Kyrgyzstan lay a three day ride away along muddy hilltops, through military checkpoints and over the Chatkal mountain range. Eagles hung in the sky, curious children clambered on our bicycles and we hurtled from the towering peaks deep into the plains of the Fergana Valley. All the way my parents were

---

[21] I rejected the ridiculous offer, obviously... and got my eyelashes dyed instead.

one step ahead, leaving rogue messages with teashop owners and chatting up the armed guards as they motored through their barricades. Puce, covered in insect bites and drowned under limp clothing, the soldiers would frown in disbelief as I confirmed that I was the person they were watching for. My mother had shown them an old picture of me and insisted we were on our way; it probably didn't help that I was buffed, bronzed and in a ball dress.

Aged 19.

Their dumbfounded shock was a pleasant distraction though. The ride east was often punishing and uncompromising; with boulder strewn roads and intermittent sustenance. We counted our blessings that Iain had brought with him extra handlebar tape and that my parents had weighed us down with sweets and biscuits.

Breakfast, lunch and dinner were a various mixture of mutton-fat, noodles filled with mutton-fat or coffee that tasted of mutton-fat.

I lived solely on a diet of Jammy Dodgers.

Uzbekistan / Kyrgyzstan Border

A maelstrom of bodies jostled for position next to the Uzbek customs building. Reminiscent of the panic of the embassies in Ashgabat, there was a lot of shouting, a bit of pushing and a bucket-load of standing in people's perspiring armpits. Not good. At the rate we were going there was no way we would make it across the border before nightfall.

"Suse—you're gonna have to tip them the wink."

"Ha ha—very funny."

"Maybe you could stand sideways..."

Usually happy to flirt us out of tricky situations, today I didn't have a hope. An hour earlier I had been stung in the face by a vicious Uzbek bee causing my right eye to swell shut and my chin to begin a distressing merger with my neckline. Half-heartedly I sidled up to the border guards and asked them some trivial questions, trying to impress upon them the urgency of our arrival in Kyrgyzstan and the worrying proximity to sunset. Unsurprisingly, none was budging. Dejected, we loitered in the warmth of the lingering afternoon, and wondered at what point my parents would incite an international incident.

But our luck was in.

A harried looking official, arms flailing in spiralling agitation, exploded through the customs building in a wave of hyperactivity. A rustle of excitement rippled through the crowd and the docile security guards spurred into frenetic activity. Someone important was arriving. Someone important was arriving right where three dishevelled Westerners had spread themselves and their grubby possessions. One of whom looked as if she had been punched in the face. Very hard. His gaze fell on us and he paused for a split second before scurrying over.

"Quick, quick. I can get you through."

We didn't need telling twice.

Kyrgyzstan

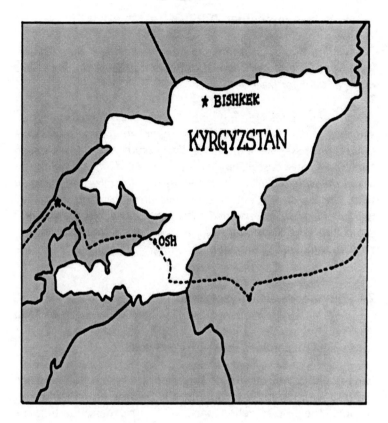

Estimated Distance: 121km
Actual Distance: 250km
Sunlight Hours per Year: 2000—3000

Riding Tunes:

"I see skies of blue, clouds of white
Bright blessed days, dark sacred nights
And I think to myself, what a wonderful world"

Wonderful World, Louis Armstrong

Kyrgyzstan

Iain's Diary—September 2009

"We've had to stay in a few one horse towns over the past few weeks and in a few of them it appears that even the horse has made a break for freedom.

After many bone jarring miles, countless punctures from smashed vodka bottles (six in one hour) and some impressive mountain passes we made it to Kyrgyzstan and a town called Osh. The Kyrgyz customs officials took great delight in telling us about the wolves which still prowl the hills in packs and going on what the insects of Central Asia have done to us recently, they will be a force to be reckoned with. Jamie has a foot the size of a small ham and Susie looks like a new born kitten with an eye swollen shut, both the result of bee stings."

24 September—Osh, Kyrgyzstan
Day 133—10243.91km

"Suse. You are enjoying yourself aren't you?"

Having embarked on one of their more unusual holidays, my parents were now questioning the sanity of following us through the less-travelled parts of Central Asia. Finding me covered in welts, sweating up mountain passes with a face so wide it was in another time zone, my mother's question was tinged with a hint of irony.

"Ha ha—of course! Give me some of that gin though..."

We had meant to push off the following morning but as Iain was once again leaving us and as I'd lost all vision to the right hand side, we rescheduled our time to cake-eating, bike-checking and preparing for the onslaught of the Tian Shan Mountains.

Iain's Diary—September 2009

The Kyrgyz Republic is hilly, very hilly, in fact. A guidebook I was flicking through described 94% of the country as mountainous and the average elevation as 2,750 metres. That is over twice the height of Ben Nevis, the highest mountain in the UK. They are a traditionally nomadic people who love their horses. National sports include a rugby-polo cross where teams fight over the headless carcass of a goat. Other sports include horseback wrestling and chasing girls on horseback in an attempt to kiss them. If the man is unsuccessful the lady is allowed to beat him with her horsewhip. The ladies also get the better horse and a head start!

To 'sweep a woman off her feet' takes on a different meaning in this part of the world. Although apparently illegal since 1991, bride-napping is the traditional way a young man finds a wife and we've been told the custom is again on the upswing. It's a way of reasserting national identity and with the rising price of weddings, a sound way of slashing costs. Bizarrely parents of the kidnapped girl generally consent to the process, although it is normally staged in present times. The Kyrgyz people have a saying: 'tears on the wedding day bodes for a happy marriage'.

Although Kyrgyzstan is predominantly a Muslim country, it appears to be a far cry from many of the other Islamic parts of the world we have been through. Islam was adopted relatively late and—as vodka flows freely—the Kyrgyz seem to have limited their version of Islam to what they can fit into their saddlebags.

The Tian Shan Mountains

Vodka was certainly flowing freely in the mountain top village of Sary Tash.

Two days before, Jamie and I had begun the epic climb, clean, rested and with a full complement of vision between us. We had bid a sad farewell to Iain and left him in the dubious care of my imperious parents. It was a crisp sunny day; the perfect temperature for riding in. There were birds in the sky, well paved roads and hardly a car in sight. Children ran behind us, cows lowed in fields and old men rode donkeys in bright local costume. We took in deep gulps of fresh mountain air and made sure but steady progress up the gently sloping inclines. Hitting the town of Gulcha by mid-afternoon we sank shots of vodka with passing lorry drivers and pedalled on to pitch our tents on the mountainside.

Only after we left town, did we remember it was bandit territory.

The day grew longer as we headed deep into the heart of the ominous peaks. Where once we had looked across the open vista of foothills, now steep oppressive rock walls blocked the sunlight and cast great shadows across our pathway. Acrid fumes spewed from the trucks that roared behind us, pushing us ever closer to the steep banks of the snaking rivers edge. Wheels jarred in potholes, lungs filled with diesel and dust scoured our throats as it billowed from thundering lorry tyres. At each settlement children gave chase, shouting for money and pawing at our panniers.

20km later we pulled over. The valley had opened out enough for us to pitch our tents and we spied a small dip that would almost hide us from the convoys. Nervously we set up camp, hoping that no one would notice. In the middle of nowhere, seven hundred dollars nestled between my socks and underwear, we were dangerously exposed. Jamie set his head torch to 'stealth' mode and we ate tepid leftover dumplings in the plummeting temperatures. Unwilling to risk a camp fire, the icy tendrils of jet black night forced a retreat to our lightweight sleeping bags, where we dawdled into hesitant sleep.

Everything looked better in the morning. Undiscovered we checked our meagre possessions and went through the usual rigmarole of an early hour decamp. Find spot to wee—check, bemoan lack of coffee—check, forget to remove additional layers which you then do two miles down the road while Jamie valiantly attempts not to get annoyed—check. A few kilometres further on, the thin valley opened out again to babbling brooks, cattle-filled meadows and friendly settlements with cheerfully coloured washing lines. Basking in the warmth of the sun we found someone selling fizzy drinks from the side of their house and busied away an hour eating apples, chatting to villagers and watching puppies annoy their mother. It was a halcyon scene.

Which immediately pre-empted a trialling afternoon.

If there is one thing more dispiriting than cycling for 15 miles up hair pin bends, roads annihilated by the constant toil of struggling vehicles and gouged into chasms by incessant mudslides, it is cycling for 15 miles up hair pin bends, roads annihilated by the constant toil of struggling vehicles and gouged into chasms by incessant mudslides—which have then been covered in sand.

"The rear wheel will do most of the work when it comes to riding in the sand. But the most important part is maintaining the momentum. If you stop pedalling, your front wheel will sink. If you notice you are going to get stuck, increase your traction to the rear wheel by performing a "wheelie" and pedal at the same time. You should keep a straight path whenever you can. Avoid sudden movements or steering. Braking will make things worse and destroy your momentum."[22]

Avoid steering? Don't brake? Perform a wheelie?

Are you freaking kidding me?!

Did I mention that we were cycling along the Himalayan belt, carrying all our possessions, juggernauts ploughing into us from either direction...

"Brrrrrrrrrrrrrrrmmmmmm"

---

[22] A Guide on Cycling in Sand, Articlebase

A few metres below we saw the exploding dust trail of a motorbike ploughing its way up the switchbacks.

"Hey—check out the biker!"

Delighted to have an excuse to stop we skidded to a slow-motion halt and waited for its rider to catch us. Mikael was on a tour of Central Asia and on his way to the same town 15km over the Taldyk Pass. We arranged to meet him there and gave laborious chase along the deep ruts of his dusty wheel tracks.

The temperature nose-dived.

On the top of the mountains the clouds came in and the mist drew close. A bank of dirty lorries lined the roadside as we pedalled past the fraying prayer flags and were bitten by the stinging winds. Conditions had once again switched in a heartbeat. Three hours later, wearing every piece of clothing we possessed and frozen to within an inch of our lives, we finally made it to the local Homestay. Greeting our roommates; Mikael, two mad Austrian men, another British cyclist and two Dutch adventurers, we huddled together in the drafty kitchen.

It was the Austrians who got the shots out.

Girding ourselves for each arctic excursion to the long-drop toilet, we drank heartily, wore Russian hats and heard hilarious tales of 'vodka-based hostage situations'.

27 September—Sary Tash, Kyrgyzstan
Day 136—10422.97km

The next morning our new best friends decamped; Mikael, the Austrians and the Dutch cyclists to Tajikistan; Jamie, myself and Charles, to China.

Charles was a 24 year old who, having finished law school, decided to pedal from Kazakhstan to Beijing before making tracks into the world of work. Tall, dark and quintessentially

English, he had spent the previous two weeks negotiating the dirt roads to the north of the country, largely getting rained on and trying not to offend the old men proffering him their daughters in matrimony. Happy to go with the flow and now to have two others to tackle the Irkeshtam border, he packed up and immediately joined forces with our depleted band. I was delighted. Able to recite the names of all Jamie's cousins and tell you his favourite food at the age of seven, it was exciting to have a new person to quiz and talk rubbish with. Jamie was a little more guarded.

What if we didn't get on?

The first cracks started on the barren mountainside. Charles had a cooker. We hadn't. Since my parents were heading across to China we had decamped some of our extraneous kit into their suitcases and packed light for the mountain crossing. Quite why we thought not having a cooker on the desolate five day journey through the 4000 metre Tian Shan passes was a good idea, I am not sure. But we had. And we didn't.

That first hour out of Sary Tash was extraordinary. Air so crisp it blasted your lungs like menthol, sun glinting off white capped mountain tops, roads as smooth as Casanova. We talked, laughed and took photos of each other gliding along against the staggering Himalayan backdrop.

Then the road disappeared.

Kyrgyzstan is not a wealthy country and its roads, often in remote and inhospitable environments, are washed away annually with the onslaught of the heavy rains. Most, then, are given only enough attention to ensure the passing of slow moving trucks and the gentle trundle of donkey carts. In 2009 the road to the border was being rebuilt by the Chinese government eager to improve their access to the trade route west.

Sadly, we had arrived too soon for its completion.

We waved at the Chinese workers toiling relentlessly in the bitter weather as, an hour out of town, the beautifully paved pathway descended back to the rib-cracking medley of rubble and sand.

For weeks now Jamie had been battling with frustration at my speed on the uphills. Though our pace on the flats was broadly similar, when it started to incline, I was undoubtedly slower. It was a simmering bone of contention. He felt that I wasn't putting enough effort in, while I pointed out that telling a crimson, exhausted women, whose heart has recently made a bid for freedom through her chest cavity, you 'hardly noticed' the gradient and are 'wondering if you can do the next one without changing gear,' could lead to the sudden and upsetting removal of a much loved area of your anatomy.

Charles was even slower than me.

"Guys, can't you go any faster?"

"We're going as fast as we can." I snapped at Jamie.

"But you're talking." He snapped back.

"What's wrong with talking? There's not much else to do up here."

"If you can talk you can't be putting much effort in."

"Charles is carrying more kit than us. He has to go slower."

"Yeah—that's true—I mean—I'm actually prepared for the ride up here. I still can't believe you guys don't have a cooker..."

In truth, it was a blessed relief that we were able to hide from the arctic winds, power up the stove and make ourselves hot coffee and noodles. Jamie begrudgingly accepted the warming fare but insisted we could have managed the mountain passes without it. Technically true, this was, though, another bone of contention.

In Uzbekistan, Zani and my parents had left us with an abundance of delicious supplies; rich chocolate, strong cheese and a bottle of throat warming brandy. Each time we stopped, I opened my panniers, dragged out some sumptuous items and chose something new for us to feast on. While before we had been splitting these two ways, now I split them three. Jamie was not amused.

"We've been on the road for four months eating mutton fat and offal. Charles left two weeks ago. Why are you giving him all our chocolate?"

"Are you seriously going to eat a Mars bar in front of him and not offer him any?"

"Yes..."

"Well, I don't mind. So you tell him. Otherwise, I'm sharing it."

"I'm not going to say anything!"

"There you go then."

Danger.

Danger is the other thing I remember about the ride that day.

Getting to the top of the mountain was an arduous, unrelenting and juddering slog; slow depressing riding in relentlessly bitter conditions. Every pothole tore through our bodies and wrenched our arms, while frozen fingers clung to the brakes and teardrops streamed from our wincing eyes. We bounded off rocks into the path of lorries, skidded on interminable grit and debris and plunged into crater after crater, gasping for the thinning air. Sheltering for warmth in a deep hollow at the summit, we drank random shots of brandy to gird us for the descent and the frightening certainty that our unresponsive fingers would struggle in their vital role of brake application. In honesty, it might not have been the safest idea to get tipsy on powerful alcohol but the lightheaded explosion almost certainly made the speed, rock and pothole combination a lot more entertaining.

Red cheeked, sooty-faced and permanently winded from the displacement of internal organs, we hit our first military check point some hours later. The Kyrgyz guards grinned broadly as they pointed their guns in our direction.

"I love you!"

The commander greeted Jamie vigorously with the only English he knew.

Irkeshtam Pass, Kyrgyzstan

Dusk was setting as we rolled into Irkeshtam. Dark eyes followed us from the shadows as we passed the burnt out shells of long rusted vehicles and mangy dogs bore their teeth as they gave chase, dragging their saggy teats beneath them. With logs blocking the road to China we wove our way through the detritus of exhausted streets to a rusty portacabin behind the tin shack of the customs building. The hotel was next door, nicotine stained walls smelling faintly of urine.

"Erm... camping?"

"Jamie, you're kidding right!"

It was bitterly cold frontier territory. We returned to the road in search of other options, finally descending on the welcoming lights of the settlement's only restaurant. To my abject relief they had rooms out the back which, though not the cleanest, didn't look like anything had died in them too recently. We were busied in by a ruddy-faced woman and set about the awkward task of unpacking three heavily laden bikes into a tiny space whilst wearing everything except our swimming costumes.

"Man I stink. I wish there was water."

"Ha—Susie, we haven't stayed anywhere with a toilet for three days. I'm not sure you're getting a shower!" Charles added helpfully.

"Baby wipe in the arm pit again then."

"Yeah—guess so. Where do we go to the toilet though?"

It was round the back of our room.

We edged out into the cold, one by one, trying to dodge the faeces of our neighbours and to block out the sound of a short but intimate liaison taking place amongst the rubble.

I awoke the next morning itchy and famished.

Dinner the night before had hardly hit the sides. We had collapsed into our lumpy beds with painful bellies, ravaged by the ruthless gnawing. I had pulled the matted blanket over the top of my sleeping bag and tried not to think about the other unwashed travellers who'd taken refuge there. Sleep had been slow and fitful, unaided by the paranoid certainty that bed bugs were crawling all over me.

At least it was only a two minute stroll to the customs office.

On the dot of nine, we lined up amongst the men in thick padded jackets scrambling to reach the counter and were jostled to the front by a weathered looking man dressed entirely in khaki. The soldier behind the desk gave us a quick once over, checked our names and slammed his stamp down on our passports. Within minutes we were back on the bikes and sailing through the barrier, shouts and whistles ringing behind us, sunshine warming our backs. We cruised the four miles through no-man's land and chattered nervously about the prospect of getting into China.

To our surprise, the next checkpoint was still Kyrgyz.

To our even greater surprise, the men prowling towards us were not smiling, rather more, 'pointing angrily and brandishing weapons'.

"Go back."

"What? Why?"

"Customs. Go back."

Determined and stony faced they waved in the direction we had just arrived. Confused, worried and annoyed that we now had four miles to pedal back up a hill we had just pedalled down, we had no choice but to turn around.

China Visa Application Service Centre

"(vi) Applicants applying for special tour to China should provide a visa notification form issued by China National Tourism Administration or the tourism administration of related province, autonomous region or municipality directly under the Central Government. Special tour includes self-driving, hiking, cycling, horse-riding, hot air balloon, or other expeditions."

Irkeshtam Pass Take Two

If we are getting all technical about it, you aren't really allowed into China on a bike. Even with a visa, if you want to ride there you should be on an organised tour with a detailed route, government guide and the properly accredited documentation. Forced to have support in Libya and Turkmenistan, there was simply no way that our budget could stretch to a Chinese babysitter for the time it would take us to cross the country. Brief investigations showed that previous cyclists had made it across the border, or snuck onto trucks and sailed over it as passengers. This was more than enough reason for us to take the executive decision of winging it.

What we had not expected, though, was to get turned away even before getting there.

A number of men came down the hill to meet us as we pedalled back towards the Kyrgyz customs buildings. They were animated, barking orders and gesticulating furiously. It didn't look good.

"We called but you didn't stop."

It had been the military, not the truckers, trying to get our attention as we left.

"I'm sorry. We didn't hear you."

I made placatory noises while Jamie bristled and Charles delved into his panniers to work on a predominantly map-based diversion tactic.

It was tense.

For one of the few times in our journey, I was afraid.

"Give me passports!"

We rustled around in our bags and begrudgingly handed them over. If the strained and ferocious looking border police decided to detain us, there was absolutely nothing we could do about it. The man in charge sniffed loudly and then began flicking through the pages, deliberately pacing clockwise around us. It was an anxious 15 minutes and an angry reprimand still before he lowered his shoulders and dismissed us with a grunt.

With a lot more trepidation, we got back on the bikes.

It was the first aggression we had encountered at a border post, and we were heading straight for another that we shouldn't be crossing.

China

Estimated Distance: 4657km
Actual Distance: 5552km
Hours of Sunlight per Year West: 2000—3000
Hours of Sunlight per Year East: 1000—2000

Riding Tunes:

"We don't need no thought control"

Another Brick in the Wall, Pink Floyd

28 September—Chinese border

The soldiers ran towards the truck and hit the deck in an explosion of chalky dust. In unison they dropped to one knee and pointed their weapons at the lorry looming large on the summit of the hill. On the ridge above, sniper guns glinted in the bright morning sunlight, loaded, waiting, trained at the cab. The vehicle was surrounded. There were shouts, orders, and then silence.

20 metres away, we stopped. Everyone stopped.

Slowly the cabin door opened and an arm emerged, reluctantly followed by its owner.

More shouting.

Awkwardly the driver edged backwards down the steps, one arm raised in surrender, the other grabbing at the railings. Black boots hit the ground before he raised both hands high above his head and hesitated, unsure whether to stay where he was or turn towards his captors. The nearest soldier sprang up and darted to him, grabbing his wrists and pushing him down against the hard bare earth. The others slowly rose from their low positions and stalked towards the truck, barrels raised.

Heartbeat
Heartbeat
Heartbeat

From our ring-side vantage point we watched as a decorated Commander pushed through the ranks and smashed the tension with a volley of barked instructions. Turning to his troops he issued a brusque nod indicating that they could lower their weapons.

Breathe
Breathe
Breathe

And—all of a sudden—the spell was broken.

Shoulders dropped, gun lowered, the soldier holding the driver helped him to his feet and patted down his trousers. The troops lying on the cliff above relaxed their guard, rose languidly and began trotting down its banks.

"It's a drill! It's just a drill." I sighed. "Oh my God. What is everyone trying to do to me this morning? I'm gonna have a heart attack!"

We pushed our bikes the final metres to the now milling soldiers where a kindly looking official looked us up, looked us down, and then smiled. Pausing only to rip the Taiwan pages from our guide book and check the images on our camera, we were whisked through the customs hall by a series of officious border police and deposited in the muddy courtyard of the isolated customs compound.

"Ha ha—man—they totally saw us coming."

As we reloaded our bags, a tour group crossing over to Kyrgyzstan drove up to the building and relinquished its wide-eyed passengers.

"Jeez Mary, look at these guys. And I thought we were adventurous!"

Suspiciously timed military exercises aside, we had made it into China without a hitch and now stood high atop the mountainous flanks of the Taklamakan Desert.

The desert mythically known as: "Go in and you will never come out", "The point of No Return" or "The Desert of Death".

Hmmm.

Set to plunge into its depths we also had other things to worry about. Rioting in the northern towns of the Western province had prompted a government crackdown across the region and once again our communications were blocked.

The vast province of Xinxiang is home to the Uighar population, a Muslim group who lays claim to the region in opposition to the Han Chinese Government. Much as the Tibetans, the local people want autonomy, yet without a global figurehead or blockbuster film behind their cause, their plight is not so widely known.

The Chinese government, keen to keep things that way, had placed an immediate media blackout across its 1.6 million km^2 in the wake of insurgence in July 2009.

We had arrived in China the week the Communist Party was celebrating 60 years in power.

More unrest was expected and the armed forces were on high alert.

29 September—Kashgar, China
Day 138—10731.24

A day after the events at the border, we hit Kashgar.

The journey down from the Tian Shan ranges had been astounding. With tarmac back underneath us we had ripped through the descent, darting between magnificent peaks, speeding past giant boulders and racing each other along deserted roads. We stopped for huge steaming bowls of spicy noodles, paddled in gurgling streams and took pictures of shaggy coated camels chewing the cud by the roadside. Though the high military presence had been a shock, we made camp loudly unafraid of thieves or bandits.

Kashgar was bigger than expected. We knew that China was not shoddy in terms of populace but coming from settlements pushing 50, it was still a surprise to find the city's wide boulevards teemed with electric bicycles and were flanked by high rise buildings.

Thankfully Jamie navigated easily through the myriad streets to a local hotel where we decamped our kit, scrubbed the grime from our bodies and went in search of as much food as we could shovel into our mouths.

Chinese Menu—English Translation

Cram food into one's face stir frying flesh

The palace quick fries diced chicken

Local explodes sheep waist

Onion explodes sheep waist

Why burn the sheep waist

Onion explodes the blocks of all

Clearly we went for item one.

First hotel, China

Kashgar was also to be the last time we would see my parents. Having been jolted over the muddy roads of Kyrgyzstan, they had spent a few days driving at the tourist restricted 50kph around Xingxang and were finally on their way back home. That night I crashed on the floor of their far superior lodgings before wandering the dark streets along to ours. Needless to say, I didn't have long to mull over the sadness of my folks departure.

"Er—guys... what have you done?"

I found the slow peaceful check-in filled with an animated receptionist and two very confused young men. The only word Jamie, Charles or I could pick up was: police. A kind Indian trader came to our rescue with a translation. The authorities had arrived and wanted us to change hotels. Much to the embarrassment of the hotel staff, we were now persona non grata and were moved under the watchful eye of the local constabulary.

The hotel round the corner, China

"Do you think it's bugged?" I asked excitedly.

"Oo—good point. I don't know. They did bring us straight to a three person room even though we might have wanted two." Jamie pondered.

"It's totally bugged.... This could be fun!"

We spent the next day consuming even more food, checking out tanks on street corners and fabricating wild and fanciful tales in the hope that someone was listening to us. We played pool in underground bunkers, got thrown out of internet cafes by the Red Guard and were generally trailed by police officers.

"No cycle in town."

"Oh no. Definitely not. We're just going to put our bikes on a bus."

At 5am we got up under the dark cover of 'official'[23] morning and hit the road.

Coming downhill into Kashgar we had been carefree and ecstatic to make it to civilisation. Now, desperate to escape without getting caught, we sidled up to road blocks and hid

---

[23] Xingxang has 'local' and 'Beijing' time. Though Beijing is thousands of kilometres away, China only has one official time, so government workers and school children arise pre-dawn and begin their days in the dark.

behind lorries, edging past them in the bustle of motorbikes and pedicabs. No one raised an eyebrow. Different police. Different orders.

By mid-morning the adrenaline of our getaway began to wane and a gentle peace settled across the team. We had left behind the damp grey streets, identical office blocks and mish-mash of cars, lorries, bikes, trikes and carts and were, as usual, back in the desert. There was nothing between us and the distant horizon; nothing but each other, the road trains and the red and white highway markers measuring distance.

We were 3567km from something.

But had no idea what.

Somewhere on the road, China

"So you're sure it's famous." Charles asked Jamie for the fifteenth time.

"Yes, well, kind of. You'll know it."

"And it's human. But not a human."

"Yeah, kind of…. do these count as questions? Cos if they do you're on 48."

There wasn't much to do in the desert. Every few hundred kilometres we would stumble across a roadside stall selling noodles and after much pointing and mimicry would eat whatever emerged and dribble it down our chins. Back on the bikes we'd wave at truck drivers, smile at police cars and draw vast, curious crowds everywhere we stopped. Maps were shown, hands were shaken and kit was repeatedly demonstrated. But a lot of the time the riding was monotonous, the journey mind-numbing and the hours painfully long.

Not as practiced and carrying a greater load, Charles was struggling with the heat and cursing the unforgiving distances. Jamie and I would take it in turns upfront, battling the headwind to maintain our speed while Charles doggedly ploughed on in our slipstream. Jamie would point out that we were bearing the brunt of the weather conditions and Charles would refer back to 'Cookergate' on the top of the Irkeshtam Pass.

Guys...

Sniping aside though, we had settled into an easy rhythm and the tensions of the previous few days had melted away. Jamie and Charles had bonded over a mutual love of food and beer and soon it was me who was the odd one out. Where I would happily lie out in the sun and wasn't fussed about what we ate, Jamie now had a co-conspirator with whom to sink Tsingtao and order the entire contents of a restaurant menu. With good food, warm temperatures and level highways, we idled away the hours listening to music and playing 20 questions as we rode.

"Want me to tell you Charles?"

"You might have to. It's been an hour."

"Ok.... Jeremy Beadle's hand."

"What? What?!" I jumped in with an explosion of consternation. "Jamie!! You wanted us to guess 'Jeremy Beadle's hand'!?!"

"It's a thing."

"Technically it is. But. What? No way. You can't have that! Why couldn't you just have Jeremy Beadle? You can't have someone's hand!"

"Yes you can Suse. You can have anything. There is no rule against having Jeremy Beadle's hand."

"That's cheating!"

"Ha—you're just upset 'cos I won again."

"I hate you."

6 October—Kuqa, China

Day 145—11466.97km

Five days, 736 kilometres and twelve super noodles later, we reached Kuqa. Jamie and I were in sore need of a rest and Charles was on his knees; time then, to find a hotel. We dragged out the guide book and made our way to a massive compound on the near side of town. Somewhat bedraggled, Jamie went in while Charles and I stayed outside to keep an eye on the bikes. Several minutes later he returned.

"Apparently they don't have room."

"What? But this place is massive. And there are only three cars in the car park!" Charles was unimpressed.

"I know. I guess they don't like the look of us."

Disheartened, we rode around.

We were turned away from the next hotel.

And the next.

And the next.

By this point it was dawning that, even though we looked like we'd spent a couple of years on the streets, it was not actually this that was stopping us getting a room.

No one was allowed to give us one.

By the fifth hotel, dusk had fallen. We were shattered, bored and increasingly despondent. Charles and I had given up hope and would have headed out of town to camp if Jamie had not been so adamant. The receptionist had been a kind and helpful lady who had tried to check us in before her manager arrived. She had also taken a bit of a shine to Jamie who, still refusing to back down, was throwing a landslide of charm at the situation. Cheekiest grin on overdrive, he happily showed her where we had come from and where we were going to on the map; simultaneously attempting random phrases from the vocabulary book.

Eventually he came back outside where Charles and I were waiting with the bikes.

"Ok—she's still saying no but let's pretend we don't understand and think she's said yes."

"Ha. Brilliant! Since it's the only plan we have. I'm totally in." I eagerly agreed.

Charles stayed with the kit while Jamie and I staked out the lobby, patiently smiling at the receptionist who, since we clearly had not left, now had no idea what to do with us.

The manager came back.

We smiled at her too.

Then waited some more.

There was discussion.

And a phone call.

Another lady arrived.

More discussion.

We grinned like maniacs as the two scurried off and the receptionist beckoned to Jamie.

"We try to help."

Another hour later, help arrived.

The police officer gave us the once over, checked our passports and signed us in as his charge. He seemed a kind, relaxed man; a kind, relaxed man very keen on the lady who'd arrived with the manager. We left the pair merrily flirting in the lobby as we lugged our bags indoors.

Excellent.

Winners all round.

An hour later we were in bed, guzzling Snickers bars and watching Travolta in drag belting out 'Hairspray' numbers on the cigarette burnt television.

"Why does this seem weirder than everything else that's happened today...?"

The Beginners Guide to Preparing For a Cycling Trip

Step 1: Get a spaghetti-strainer and several small sponges. Soak the sponges in salt-water and paste them to the inside of the spaghetti-strainer. Place the strainer on your head. Find a busy road. Stand by the side of the road and do deep knee-bends for eight hours. This will acclimatize you to a day's ride.

Step 2: Take some sandpaper and rub your rear-end and the insides of your legs for about 20 minutes. Rinse with salt-water. Repeat. Then, sit on a softball for eight hours. Do this daily.

Step 3: Each day, take two twenty-dollar bills and tear them into small pieces. Place the pieces on a dinner-plate, douse them with lighter fluid and burn them. Inhale the smoke (simulating car-

fumes). Rub the ashes on your face. Then go to the local motel and ask them for a room.

Step 4: Take a 1-quart plastic bottle. Fill it from the utility sink of a local gas station (where the mechanics wash their hands). Let the bottle sit in the sun for two or three hours until it's good and tepid. Seal the bottle up (kinda, sorta) and drag it through a ditch or swamp. Walk to a busy road. Place your spaghetti-strainer on your head and drink the swill-water from the bottle while doing deep knee-bends along the side of the road.

Step 5: Get some of those Dutch wooden shoes. Coat the bottoms with gear-oil. Go to the local supermarket (preferably one with tile floors). Put the oil-coated, wooden shoes on your feet and go shopping.

Step 6: Think of a song from the 1980's that you really hated. Buy the CD and play 20 seconds of that song over and over and over for about six hours. Do more deep knee-bends.

Step 7: Hill training: Do your deep knee-bends for about four hours with the salt-soaked spaghetti-strainer on your head, while you drink the warm swill-water and listen to the 80's song over and over. At the end of four hours, climb onto the hood of a friend's car and have him drive like a lunatic down the twistiest road in the area while you hang on for dear life.

Step 8: Humiliation training: Wash your car and wipe it down with a chamois-cloth. Make sure you get a healthy amount of residual soap and road-grit embedded in the chamois. Put the chamois on your body like a loin-cloth, then wrap your thighs and middle-section with cellophane. Make sure it's really snug. Paint yourself from the waist down with black latex paint. Cut an onion in half and rub it into your arm-pits. Put on a brightly coloured shirt and your Dutch oil-coated wooden shoes and go shopping at a crowded local mall.

Step 9: Foul weather training: Take everything that's important to you, pack it in a Nylon bag and place it in the shower. Get in the shower with it. Run the water from hot to cold. Get out and *without drying off, go to the local convenience store. Leave the wet, important stuff on the sidewalk. Go inside and buy $10 worth of Gatorade and Fig Newtons.*

Step 10: Headwinds training: Buy a huge map of the entire country. Spread it in front of you. Have a friend hold a hair-dryer in your face. Stick your feet in toffee and try to pull your knees to your chest while your friend tries to shove you into a ditch or into traffic with his free hand. Every 20 minutes or so, look at the huge map and marvel at the fact that you have gone nowhere after so much hard work and suffering. Fold the map in front of a window-fan set to "High".[24]

Little things

While we were braving the vast expanses of Western China, my parents had got news to Iain about the communications vacuum and he was helping us update the SolarCycle website. Having pilfered the words above from the website of 40,000 mile cycling veteran, Al Humphrey's, he had taken the opportunity to describe the situation we would be in to the folks back home. On the road, I was thinking of some other words Al had given me before we set off.

In 2007 he had cycled across Siberia with a friend.

In a move that I still cannot begin to understand, they had decided to attempt this, time constrained, in the heart of winter. But it was not the distance or freezing conditions that annoyed Al, nor the near impossibility of sleeping in minus 40°, it was the way that his friend chewed his food.

"It just got to me. He had this really strange way of eating. It made me want to scream. In the end I mentioned it and he just stopped. It was such a small thing. But sometimes, living that close to someone under such extreme conditions, it can be the smallest things you need to talk about."

---

[24] Unknown—but insightful—cyclist, via the website of Al Humphreys

An unexpected military welcome. Western China

Charles eating a snack but dreaming of this chicken... China

The road through the Taklamakan desert. Otherwise known as 'the desert of no return'. Western China

Jamie and Charles taking a pit stop – with Charles' cooker, Western China

Building the highway East—some of the many roadworks we came across in the Taklamakan desert. Western China

A monk takes a good look at our kit. On the Tibetan Plateau, China

Spot the world's smallest snowman. On the Tibetan Plateau, China

Though the snow wasn't the best thing for the bikes! China

A long old way. China

The world's biggest PV panel on a building—all 6900m^2 of it. Suntech, Wuxi, China

Diner breakfasts and highway patrol men with guns. Western USA

Always a new challenge. Puncture from the inside! USA

Battered by rainstorms from the Gulf of Mexico, USA

"Orlando. Stop it. This is getting embarrassing..." USA

Tsk! USA

Round the world and round the Sebring track—time for a spot of celebrating. USA

Jamie's Diary—October 2009

Susie's alarm went off at 7 and I finally made myself get up at about 20 past. Decamped and faffed around with the heart rate monitor and phone batteries for a while. It was chilly so I tried my new gloves for the first time. Susie almost lost a tent peg but managed to find it in the dirt after scratching around for a while. This surprised me as I'd expected her to give up fairly quickly. We left the campsite at around nine.

Susie's—response, on reading Jamie's Diary Entry:

JAMIE'S DIARY ENTRY:

"Susie almost lost a tent peg but managed to find it in the dirt after scratching around for a while. This surprised me as I'd expected her to give up fairly quickly."

COULD READ:

"One of Susie's tent pegs came out in the night. Cleverly she carefully marked the spot in which it lay so that she could pack her tent up the dark the following morning but find it when the sun rose—ensuring no delay to our departure. Despite this, I had a go at her twice for being stupid for taking her tent down before finding it. In retrospect it amazes me that she didn't bite my head off. Especially as I have lost all of my tent pegs and she has given me some of hers. As I borrowed some off Charles too I now have more tent pegs than she has but I don't think she has noticed. She can't put her tent up as well as I can as she has to put several ropes on one peg. This means I get to ridicule her each time she tries."

Catty!

Thank goodness Charles was there.

9 October—Korla, China

The blog-based tent peg showdown, almost 150 days into the trip, was one of the first times that Jamie and I had actually bickered. Both fairly relaxed people, we would snap occasionally, but until now had managed to swerve a confrontation. Jamie was frustrated that I would not listen to him; while I was annoyed he'd assume I couldn't do something.

Clearly, he was telling me what to do.

And I was ignoring him on purpose.

All in all though, it was a happy band that rolled into the petroleum town of Korla. Having spent the five days since Kuqa in near isolation, we were shell-shocked to find just how huge and how wealthy it was compared to the nothingness surrounding its borders. Other than camels and a few jovial truck drivers, we'd passed little more than the odd patrol point, an infrequent village and signs to the illusive 'Thousand Buddha Caves'. We'd played 'guess the road sign', 'guess where the road sign might take us' and 'guess if we can use the road sign as something to camp behind' before pitching our tents under the desert stars. This industrial town was a far cry even from Kashgar.

Shiny new buildings lined its wide thorough fares and flashing traffic lights lured us down its busiest streets. Welcomed into a ridiculously cheap and opulent hotel, five staff lined up in depreciating height order, as we heaved our way across the lobby. We wondered what it might have been near Kuqa that brought about our curt reception; Gulags, nuclear sites, Uighar revolutionaries? In a country that jails political activists and celebrates anti-drugs day by executing addicts[25], we thought it best not to ask. Instead we washed, pruned and filled our shrinking bellies before navigating the marble staircase to our clean pressed beds.

---

[25] Erm – I'm not sure that's what the UN intended...

The night before we'd slept in a drainage ditch.

Somewhere in the desert, China

Back out on the road things were not going so smoothly. Though much of the riding had been along unopened motorways with smiling workmen and freshly laid tarmac, the ride into Turpan was proving a challenge.

Quite literally.

As I could not pedal.

After a night's respite we had left the sanctuary of our luxurious abode and set off for another three days of labour. The roads had led through the hills to the North of the Taklamakan and the difficult terrain had been followed by an exhausting change in conditions.

Headwinds.

Before these had been tempered by buildings, trees and vehicles but with nothing between us and the epic forces, the gale crescendoed into terminal velocity as it raced across the desert floor. We had come across headwinds before. Mentally you settle in for a long, weary slog dreaming of the moment the gusts will dip and you can propel yourself on without a strength sapping struggle. 1,400km into our desert crossing, though, struggle was not even an option. The wind was simply too powerful for me to push against.

"This is insane Jamie!"

"What?"

"Insane!"

"I can't hear you!"

Jamie valiantly came to my rescue, sheltering me much as possible from the brunt of the storm by riding in front. Even so, he could not stop the huge blast that threw me off the road and down the gravel bank.

"That was a complete 360°. Your bike went over your head!"

With no injuries sustained it was a surprisingly pleasant interlude to the tedious slog of our slowest miles.

Fortunately the unstoppable winds that hurled themselves across the Takalaman proved a veritable boon as we turned the corner and headed east. We flew past giant wind farms, rested by random sculptures and stung our lungs on fields of chilli drying by the roadside. We stopped in Jiaohe, the ancient sandstone city, and drank rancid wine by its weather worn stupa field. We bought new screws for our solar panels, patches for our punctures and clunked through a variety of ill-fitting bike chains. Replacements bought from kerbside mechanics would inevitably work until it was too late to change them and we'd nod through the gentle 'ker-chunk' of movement until we chanced across another town.

Minutes limped by into hours, and hours limped by into days; all filled with monotonous expanses, tasteless luncheon meat sausages and ever more unfeasible ideas for a film script revolving around ninja assassins.

By now we were able to ride lengthy distances without issue, set up camp in double time and vaguely comprehend the Chinese road signs. We kept ourselves entertained through the relentless grind of not-very-much-at-all by trailing tractors, daring Charles to put extra spice in his noodles and doing everything in the style of Jackie Chan.

All too soon we had crossed the 900 miles to Jiayuguan and emerged by the fort at the very end of the Great Wall of China.

Susie's Diary—October 2009

The Jiayuguan fort signifies the point at which, traditionally, Chinese civilisation ended. Those cast from the land spent their last days there and it was the final staging point on the silk route west. It is a quintessentially oriental collection of buildings from which the Great Wall radiates. Built to keep out the 'Hun' in the North, the Wall (or several walls) is 25000km long and was, during the Ming dynasty, guarded by a million soldiers.

I can confirm that it is currently being guarded by an evil dog with horrible red eyes that pulled its chain from the wall and ran at me until I tried to twat it on the head with a bicycle pump!

20 October—Jiayuguan, China

Day 159—13187km

We had made it. We were through the Takalaman, out of Xingyang and back on-line. I turned my phone on, awaiting the onslaught of messages. Beep. Beep. Beep.

Three? Only three? Ah well, at least a couple of people care.

Message 1

"Half price haircuts at Rush. Just quote 'Rush Offer' before 1st November."

Message 2

"Hey Everyone, just a reminder of the address for my house party..."

Message 3

"Hi Suse, you still in the desert? Everything ok? Lots going on here... Hope you're having a great time. Shenaz"

Or just my mate Shenaz!

Still, it was two more messages than either Charles or Jamie.

Though this time we had been able to warn friends and family about our lack of communication, it had caused another problem. We had failed to organise our visit to, and subsequently missed the turn for, two solar stations.

Too ill to make the most of the concentrating solar power station visit in Egypt and with breaks in tracking undermining the impact of our solar kit, I was worried that we had missed another chance to highlight the aims of our expedition. With the Chinese CSP stations and our opportunity to profile its deserts 600km in the direction we'd just come, though, there was no persuading either of the guys to turn around.

"Susie. Seriously, we've cycled 100 miles a day since Iran— through the Tian Shan Mountains and the Western Chinese desert. No one is expecting an article on solar power!"

In honesty, knowing those 600km would be into the winds that had just pushed us to Jiayuguan, I was happy the decision was taken from me.

Feeling guilty, disappointed, yet a little bit relieved, we cut our losses and carried on.

22 October—Zhangye, China
Day 161—13443km

Fort visited, coats purchased and socks washed, our rest in Jiayuguan was only brief.

From the battlements of the garrison we had looked back across the plains and thought of all those who had once been thrown out into them. Cast from the realm of the Chinese Empire, the ne'er-do-wells of ancient times would traverse the miles to Samarkand by foot, hoping to find water and praying for their lives. They might chance upon a merchant or trader with whom they could barter a price for their passage or be lost to the sands and the mountains forever. We had crossed the Taklamakan and cleared the passes, but we'd done so with far greater chance of survival. We had two wheels, two panniers and the definite advantage of a largely tarmacked road. But it had still been a relentless grind and we weren't out of the desert yet.

Two days after we reached the Wall, built to keep out the marauding Hun, we lugged our grubby panniers onto our bicycles and saddled up and back to the fray.

At our first pit stop—and following a spirited attempt to make Jamie eat a snack pack of chickens' feet—we idly assessed the road ahead. Conditions were good, the road was smooth and the gentle terrain seemed pleasantly manageable. Five minutes later and our excitable brains were coaxing more reluctant limbs into an epic mission to Zhangye, 220km further. With the commanding winds still in our favour we had a fighting chance of making it by night fall, which would not only allow us a quick trip past a massive Buddha, but mean we'd be out of the sands forever...

Hours later, cycling the wrong way up the hard shoulder of the pitch black motorway, juggernauts pelting past, horns blaring in horror and aggression, it occurred to us that a 246km escapade might not have been the best idea yet. In fairness to Jamie, about 170km in he had begun to protest due to safety. But Charles had been keen for a spot of sightseeing and I was adamant that no amount of suffering could outweigh the blissful notion of a shower two days in a row.

Naturally, it was now nearly midnight and we had not only missed the exit to our destination (well, I hadn't, I had seen it but was told I was wrong and then ignored, again, not that I'm still bitter or anything) but we had run out of food, water and the will to continue.

Damnations.

The next morning having finally totalled 256km[26], fully loaded, through hills and exhaustion, we ached our way out of our budget hostel and headed straight for the swankiest establishment in town. With the guys appetites whetted by the toil of our foolish endeavour, the level of consumption at the buffet breakfast was almost breath taking, and it was a three hour eating frenzy, and subsequent slumber in the Buddhist sanctuary, before we were finally ready to hit the road.

At which point we promptly did something even more foolish than the day before.

This time it was nothing to do with clambering into a car with a man wielding a machete or the random decision to cycle twice as far as usual. It was a more generic piece of uselessness, a total avoidance of common sense, a glance at reason before a quick 180° in the opposite direction.

We took a short cut.

Without assessing that the 'short cut' was via the Tibetan Plateau, otherwise known as 'the highest place on Earth'.

Susie's Diary—October 2009

Well, that was unexpected...

The beauty of undertaking a trip to highlight the power of the sun and the deserts is that it is usually warm. If not utterly sweltering.

Man alive—there was a moment out there when Charles' beard froze and Jamie nearly crashed as he couldn't feel his hands to work the brakes!

---

[26] Ten extra for time spent travelling in the wrong direction.

Short cut, China

The late start from Zhangye coupled with tired legs led to a short 65km day but one that was all uphill. The cold bore down on us as we hit the town of Milne and we headed to bed filled with trepidation about the mountains ahead. We were climbing into the snow line and were woefully underprepared for it.

The next day's ride was very slow going. But I maintained a constant drumming on my handlebars to keep the circulation in my fingers and we made the first 3685 metre peak in the midday sun, buoyed by Snickers and a great sense of achievement. Jamie built a snowman, we ate a picnic and then busied ourselves hurling snowballs at whoever attempted to go for a wee. Speeding down the other side of the mountain, the icy sting of the wind was a quick sharp shock, but it took our breath away only marginally more than the hairy yaks, majestic peaks and the monks on motorbikes, orange robes flowing.

By the top of the next 3765 metre peak, though, the mood had changed. The clouds had come in blocking out the late afternoon sunlight and our legs were giving way underneath us. Wrapping our leaking, semi-iced noses under the cover of makeshift balaclavas, we plummeted back down to the heart of the valley, debating whether to go slow and minimise the agony or to go fast and get it over with. When we eventually began to churn through the uphills our thighs cried out in residual agony, but it was a relief to have warmth return to our bodies. Just before dusk we scanned the horizon and saw in the distance the town we were aiming for.

All two houses of it.

We broke into a shed, used our Space Blankets to block out the torch light and settled in for a restless night under the crashing temperatures. Rising early to avoid detection, the next few hours were our coldest yet. This was the 'ice on beard and total loss of feeling in extremities' moment mentioned in my diary entry. We were saved at the nearest town by a wonderful man who let us sit by his hearth and use it to defrost our fingers. He was certain our plan to cycle to Xining, over yet another high pass, was folly.

Naturally we set off regardless.

As it turned out, the highest pass yet was not as terrible as we imagined. The heat of the day saved us from the glacial conditions and the gradient of the incline overheated us as we rode. We trundled through small villages with smoke drifting from chimney tops and looked across at alpine meadows dotted all over with grazing cattle. Sweet, clean air filled our lungs and we settled into a steady rhythm, plodding around the winding hairpins. Some monks stopped to look at our bikes at the summit and we boiled up some coffee in the midst of the snowline. The descent too was nowhere near as awful as feared, though my eyes still were still streaming and Jamie insisted his hands were frozen.

Though only because I said that, if he couldn't feel anything, he could warm them by sticking them up my T-shirt[27].

Tibetan Plateau

The conditions really were harsh. But the scenery majestic and staggeringly beautiful; jagged snow-capped mountains, herds of wild animals and smiling, happy, ruddy faces, beaming out from thick woollen overcoats. Additionally, since we'd missed our solar date in the desert, heading to the region at least gave us something to write home about.

The Tibetan Plateau is a vast store of ice and the closest point on earth to the sun. When we had finally made up past the onslaught of switchbacks, we had got off our bikes, taken a long, deep breath and looked at the view from the roof of the world.

---

[27] You just have no idea what you might suggest one day in order to get someone to STOP THEIR WHINING.

"The Tibetan plateau gets a lot less attention than the Arctic or Antarctic, but after them it is Earth's largest store of ice. And the store is melting fast. In the past half-century, 82% of the plateau's glaciers have retreated. In the past decade, 10% of its permafrost has degraded. As the changes continue, or even accelerate, their effects will resonate far beyond the isolated plateau, changing the water supply for billions of people and altering the atmospheric circulation over half the planet."[28]

"This region has a near inexhaustible source of solar energy due to its average annual radiation intensity of 6000–8000 MJ/m², ranking it first in China and second after the Sahara worldwide. Currently, Tibet has 400 photovoltaic power stations with a total capacity of nearly 9 MW. In addition, 260,000 solar energy stoves, passive solar house heating covering 3 million square meters, and 400,000 m² of passive solar water heaters are currently in use in Tibet."[29]

26 October—Xining, China

Day 165—13793km

After a long and winding descent, strewn with ramshackle houses and red-gold monasteries, we slammed straight into the hustle and bustle of Xining, a huge, modern and neon clad city. It was the second time the country had given us a culture shock. People talk about 'two Chinas' but this was our first introduction to the populated half. Newly built and housing millions, the buildings were grand, the expressway imposing and the streets were littered with iridescent billboards. We rode past, mouths aghast at the bright fluorescent advertisements for shampoo, teen pop idols and our much loved 'Sod' apples.

Modern China was, though, to treat us as kindly as its outposts. Our first meal was generously discounted as the owner liked foreigners, and we took a quick side trip to Ta-er Si, one of the

---

[28] Physics Today
[29] Science Direct

most important monasteries outside Tibet, birthplace of Tsongkhapa and former home of the current Dalai Lama. The city also presented us with a much needed opportunity to launch ourselves back into consumer society.

Shopping list:

Shoes—flip flops just don't cut it in the snow
Gloves—for Jamie, so no more excuses
Snickers—it is hard to express our love and devotion to the only international chocolate to infiltrate China's heartland

It was also time to take stock and to give ourselves, and our bikes, a bit of a once over.

Bikes: muddy, clunky, would probably hold out

Us: muddy, scrawny, would probably hold out

While we had been in the desert there had been a number of occasions when we simply couldn't get enough food to equal the energy expended. Fully loaded, riding fast, we would plough through 500—600 calories an hour and though we would eat when we could at road side restaurants, much of our food was courtesy of the intermittent motorway service stations selling nothing more than chow mein, feather-light biscuits and row upon row of sugary Pepsi.

It soon became apparent that there are only so many packets of noodles that one person can eat, and only so many cans of Pepsi they can consume, before both a serration of the liver and an extreme dental emergency. We had maxed out our life time quotas of both and were now in need of a couple of vitamins. Like some sort of food camels, the guys would eat anything they could find and keep eating and eating until normal people would have given up... or at least tactically vomited a couple of times.

I could not, and when it was offal, would not, do the same.

I looked in the mirror of our hotel room at my jutting out rib bones and jabbed curiously at my prominent hips.

Pies on the menu for me then.

Jamie, half naked, bounded in from the toilet; delight in his eyes and a grin on his face. Charles and I turned to see the cause of the commotion.

"Hey—check it out, check it out! It's a six pack."

"Ha—Jamie, you're breathing in!" Charles spluttered through a mouthful of Snickers.

"Ok, ok a four pack.... A two pack....?" He laughed, eyes still beaming.

"Are you sure you actually had one when you were at Uni, J?"

Jamie and I needed a day to organise so Charles went to Langzhou to check out some 1000 Buddha caves while we drank Tsingtao, battled sallow faced gamers for use of the internet and plotted our attack on the following months.

It was amazing to be indoors for a change.

29 October—Hong Gu, China
Day 168—13905.04km

"Jamie, why are you smirking?"

"No reason Suse."

"I'm suspicious."

"It's nothing. Honestly. The room's lovely..."

Two days after Charles had left and a load of planning, blogging and organising later, we were on our way to catch him up. We had stopped for the night in a small town by the roadside and located the only hotel.

"What? It's nice. It's fine. It's.... Jamie? Jamie? Ha ha—why is there is a window from the bedroom to the bathroom? Ah—I get it. It's one of THOSE hotels. Oh man, if you watch me wee there's gonna be trouble!"

Either this was a unique Chinese design or we had unwittingly stopped in a brothel. I assume that he averted his eyes as I went to the toilet, though the next night I know for a fact he didn't!

After a long day on the bikes, we had stopped in a truck stop and spent £1 on a room. Naturally the price reflected the facilities and the toilet was a bucket in the corner, or a hole in the floor of a shed 50 metres away.

Until now I had always gone out of the room.

But it was just so cold.

And so dark.

And—oh well—who's gonna know?

CLICK.

"B*stard! Did you just take a photo?"

October 2008—a year earlier

"Hey Suse, fancy a mini cruise?"

"By 'mini cruise' do you technically mean 36 hours on the ferry?"

"No. Kind of. Ok—Yes..."

A year previously, I had been cajoled into a trip to Spain by a friend who had a long tedious drive ahead. Aside from watching the show tunes cabaret and trying to spot dolphins, there was not a vast deal to do for two days on a boat, so we holed ourselves up in the bar and got chatting to the guy on the seat next door.

He was chivalrous, entertaining and just setting out on a cycle around France.

30 October—Langzhou
Day 169—14015.16km

Back in Langzhou I got a call from Charles.

"Hey—what's up? Everything ok?"

"Suse, you'll never guess what's happened. I've just met a guy who knows you!"

When Charles had arrived in the city two days earlier he had booked himself into a large hotel in the centre of town before heading out to pick up supplies. When he returned to the hotel he got his key from the reception and went to his room to relax for the evening. Only it wasn't his room. Looking around he recognised the tell-tale signs of panniers and over worn socks, but they were not his. He left a note on the door before heading back to the lobby.

Which is how Charles (bearded English cyclist) met Humphrey (bearded English cyclist).

Unsurprisingly the receptionist had mixed them up and given Charles Humphrey's key.

More obscurely, though, Humphrey was the guy I had met a year earlier on the ferry.

1 November—Tianshui, China

It had been a long ride into Tianshui and we were due to meet the guys on route. We had glided past terraces littered with wheat fields and ridden past vegetable sellers balancing baskets on bamboo poles. We scanned along the shop fronts as we reached the city's outskirts, past crumbling whitewashed buildings covered in faded Chinese script. As the afternoon drew closer, Jamie and I found Humphrey cross legged on the roadside. He was fixing a puncture without a care in the world and looked up at us with a massive, welcoming grin. We offered to help but he insisted he was fine, waving us on towards Charles a few miles further. We found Charles in a grubby restaurant that looked out across a metalworks and settled into cold steel chairs that scraped from under the Formica table. Playing eeny meeny miney mo we made a stab at the menu and caught up on the news over a bottle of beer.

"Jamie, prepare yourself now, Humphrey does not travel at speed..."

Charles had enjoyed an incredible couple of days, but now used to our pace, was getting slightly frustrated with the constant stop-starts. It was a while yet before Humphrey eventually emerged, dark by the time we hit town, and late by the time we were ready to forage for food.

Desperately hungry we dumped our things in a cheap hotel room and hopped in the lift to head back downstairs.

The journey downstairs

Floor 10: Doors open—men wave trying to entice us to their restaurant
Floor 9: Doors open—office rooms, two women look up and smile
Floor 8: Doors open—massage parlour, lady shouts out for us to stop

Floor 7: Doors open—another restaurant, waiter smiles
Floor 6: Doors open—cafe, more waving
Floor 5: Doors open—15 women stand dressed as a sexy version of Little Miss Santa....

JAB, JAB, JAB, JAB, JAB

I have never seen Charles move so fast!

As the doors closed he took one look at Jamie, leapt the breadth of the lift, elbowed me out of the way and started to hammer at the control panel.

For those who have not yet done the maths, there were three men in that lift, one of whom had just pedalled through the Middle East, and all of whom had just ridden across the Western Chinese desert.

There were 15 scantily clad women on Floor 5.

JAB, JAB, JAB, JAB, JAB

Shudder

Clunk

"Oh sh*t!"

The frenetic pounding had proven a bit too much for the mechanics of the elderly pulley system and lift had come to a shuddering halt. We were stuck. Pushing back the tears of laughter I picked up the emergency phone.

It came away in my hand.

Ah.

Jamie sprang into action.

In a scene straight from the 'Man's hand guide to Die Hard 4', he ripped off his jumper and began to prise the thick metal doors apart. Muscles bulged, veins popped and a low and guttural grunting noise was violently emitted.

Then the shrieking began.

Seeing a square jawed, rugged Westerner heave the metal doors open a sliver at their feet, all name of pandemonium broke out amongst our would-be rescuers. Hands wriggled through the gap, screams of joy and excitement perforated our eardrums and cries of "We save you, we will saving you" detonated through our metal prison.

We valiantly attempted to stay calm in the face of the incredible waves of oestrogen seeping through the tiny opening and busied ourselves in the important task of taking pictures of each other stuck in the lift.

Half an hour later and we were crawling through the gap being dragged to safety by the Claus sisters, each desperately grabbing at the broad shouldered gentlemen emerging like action men right at their feet.

"Choose me! Choose me! Choose me!"

I have never seen the like. There was a lot of grasping, a lot of hugging and a lot of noises so high pitched that dogs must have come running. I am hoping it is the closest I ever get to experiencing a Justin Bieber concert.

"Hey guys, want to stop for a drink?"

"Totally. After we go and grab some food."

Which ably demonstrates the exact order in which men think.

The hills near Baoji, China

All the way across China we had been pedalling down its expressways. The Western part of the country is decidedly less developed than the East and in many areas huge road construction projects are underway to link the nation together. These expressways are marked intermittently with check points manned by local police units and a series of officials dressed in high visibility jackets. After our run in—and run away from—the police in Kashgar, each of these checkpoints brought with it a degree of foreboding.

We had met some other cyclists on the road who had fallen on the wrong side of the authorities and spent the night in jail for their troubles. But with very little ammunition in our cyclist arsenal we were almost solely reliant on the 'smile and wave'[30] technique to get us past the officials blocking our path.

On the way to Baoji we had spent a long time on small, overcrowded, potholed roads and were beginning to tire of the constant jarring and slow moving progress. Now in the industrialised part of the country the skies were a consistent grey. Diesel fumes from passing lorries clogged our lungs as we heaved our way despondently through the laborious hill country. The expressways went through the hills.

"You guys thinking what I'm thinking?"

Susie's Diary—November 2009

The entrance was text book.

Sidling up behind a lorry, the bemused security guard didn't stand a chance in the face of our frenetic waving and frantic last minute speed pedalling.

---

[30] The 'smile and wave' technique prays on the natural reflexes of the human condition—i.e. the inability not to smile and wave at someone smiling and waving at you. Before the police had time to stop us, we greet them cheerily and turned into dots on the distant horizon.

We raced off down the road. Adrenaline pumping. Praying that they would not give chase. The motorway had cut through the mountains like a scythe. Where once we were riding over broken roads through precarious hill side towns, flanked by colourful Buddhist shrines, now we were rolling past sheer walls of rock, nonchalantly gauged into the very heart of the mountains.

Some of the tunnels were over 3km long. Brightly lit, shining like beacons in the gathering dusk. We ploughed through them, conscious of their narrow lanes and praying we didn't get a puncture. The fumes were overwhelming and the earth shuddered as lorries thundered past inches from our loaded bikes. I have never heard anything as distressing as the noise of a police siren echoing 500 metres underground.

Especially when I thought it was directed at us.

Back in the hills near Baoji, China

Though it perforated our eardrums, we were relieved when the police convoy sped straight past. At the next check point, however, things did not look quite so fortunate. Doubling as a weigh station, police thronged around the toll booths and traffic was directed into channels by a flurry of officials brandishing clip boards. There was no way we were getting out of this and nowhere to go back to but the perilous, desolate mountain roads. In the deep valley the sun had nearly disappeared and our chances of surviving the winding bends on the B-road would diminish significantly if we kept pedalling. We were far from anywhere and had no supplies. We would have to camp dinnerless on the cold, muddy hillside.

BOOM!

Or would we?

As the authorities came out to greet our errant band, a truck behind us careered wildly out of control and hurtled past, smashing into the barricade in an eruption of sparks, shrapnel and screaming metal. In the face of a careering vehicle the assembled body of neon clad officials dove out of the way and hurled themselves over the low concrete barricades.

Lady luck had exploded to our rescue.

"Just keep going! Pedal, pedal, pedal..."

4 November—Xian, China
Day 174—14673.51km

Though we had made it to Baoji via the highway that night, eventually the police had caught up with us and ushered us off. In the wealthier eastern provinces the 'smile and wave' approach was thwarted fatally by the proliferation of cars and vehicles in which our pursuers could nimbly hop and merrily give chase. The graceful ease of motorway riding was now replaced with weaving along rural lanes, dodging smoke spewing ancient buses and slamming into prodigious gouges in the tarmac. It was two rib shattering days later before we finally hit the outskirts of Xian.

Culture shock three.

Although we had seen the bright neon sights of a Chinese city already, we were unprepared for quite how modernised the country would become as we crossed it. Following 20 dreary kilometres of brick works, offices and communist apartment blocks, we reached the ancient stone walls and the staggering drum tower at the heart of Xian's 3000 year old metropolis. We weaved through the traffic and snaked down back alleys, past market stalls, chickens' heads and bright glutinous rice cakes. We swerved round three wheelers, dodged wobbling bicycles and passed hoards of pedestrians before turning onto the main street: Starbucks, Gucci, Dior, Cartier, McDonalds, Baskin Robbins, Guess, Pizza Hut.....

We had a day off. I was going shopping.

CODE ORANGE

With Humphrey having defected on a visa mission of his own, Charles, Jamie and I had raced to Xian—both to see the Terracotta Army[31] and in a desperate bid to outrun the weather. With a Code Orange warning of plummeting temperatures and nightly reports of gale force winds, we had hoped to dodge the worst of the conditions and make it to town before the tempest arrived. Still pitifully underdressed and underprepared for the sub-zero temperatures, I headed out to get a jumper while the boys did some sight-seeing. They returned with some Mao-Tai[32] and enquired after the success of my expedition.

"So did you get some warm clothes Suse?"

"Not exactly, but I did get something....."

Every girl needs a pair of red six inch stiletto heels right?

Focusing a little more on the 'first night out in five months' than the arctic blast bearing down on us, we got ourselves ready and set out on the town. We had been dreaming of this moment for weeks, if not months. Though we'd had a few beers in the course of our meandering, the fairly strict schedule and route through the desert had not allowed much of a chance to make merry. Xian is a city with singing, dancing and a road it calls Bar Street...

The next morning

"Ok, so who remembers anything?" Charles was the first to brave the complicated activity of speech.

---

[31] The "Terra Cotta Warriors' are life-size sculptures of the armies of Qin Shi Huang, the first Emperor of China. The warriors were buried with the emperor to protect him in the afterlife.
[32] Exceptionally potent Chinese rice wine.

"Erghmpf. Stop talking so loudly. I think I'm blind." I mumbled, face in pillow.

"You haven't opened your eyes yet!"

"I remember leaving the room." Jamie came round and joined in. "And definitely getting to the bar. Yep, and haha!"

"What? Why are you laughing?"

"Don't you remember? Susie fell onto that table trying to show some guy her shoes."

"Oh yeah. Hahaha. I remember that too." Charles confirmed distressingly quickly.

"No. No. You don't... You're lying." I accused with faint hope.

"Ah—you wish. You did. And then you went to the toilet and fell in cos you were squatting in heels."

Charles' confirmation of my activities brought with it a newly distressing revelation.

"No. Really? No. Did you see?! How would you even know that Charles?"

"Well, you told everyone. Then stumbled around a bit. Oh and I think I got locked out of the room until you guys came back."

"Why didn't we come back together?"

"I'm not sure Jamie, but I hope it isn't suspicious that you and Susie passed out in the same bed...."

Susie's Diary—November 2009

"There comes a time in every person's life when they take stock, look themselves straight in the eye and ask:

*"What the **** am I doing?!*

Last night was one of those times.

Dusk had plummeted upon us, we had been cycling into a snowstorm for several hours and I had just fallen into a chasm in the road.

A car pulled alongside.

"We can help you?"

"No, no, all fine."

"We want help you."

"Really, no, all ok."

"Please can help you?!"

I am not sure if the passers-by had stopped because of the snow covering me, the fact that my eyelashes had iced together or because I had been nearly mown down by a lorry in the twilight snowstorm. But it was kind of them nevertheless. And as the tail-lights of their car faded into the mist, I paused to wonder if this 10km was really, actually, worth it.

We had a goal though. And that goal was Zhoukou. Right then, it was all I could think of and all I was clinging to. Well, that and the thought that Zhoukou might proffer a hotel. With hot water. And heating. And, and...

"Slam. Boof. OWCH"

Sadly, in the short spell assuring the genial motorists that we did—indeed—mean to be out on the highway-of-snow-filled-doom in the dark, ice had formed over the cleat in my cycling shoe. I pushed down hard but it had not clicked back into my pedal.

I hit the seat with force.

Then skidded a bit more for good measure.

Ar*e!

Going had been getting tough. Encouraged by the vicious headwinds, the trees had been pelting us with shards of ice while the snow and slush spraying into our faces was not only blurring our vision but causing the gears, pedals and brakes to freeze as we moved. For a good hour or so we had only been able to go at one speed, chains clunking and lurching. We were praying we would get to the navigable parts of the road before the other traffic so that we wouldn't need our unresponsive brakes.

Now I could no longer turn the wheels.

Normally, not being able to 'clip in' to my pedal is not an issue. I simply swing it round 180° and use the regular side suited to normal shoes. This time, though, with ice covering both sides of the metal, it was a battle to gain traction. The problem was amplified by the fact I had absolutely no feeling in my foot and no idea when it actually met its target.

Clunking, slipping and now one-legged, I limped the last few kilometres to town."

11 November—Luoyang, China

Day 181—15077.54km

For some days now it had been getting colder.

A little rough around the edges, we had left Xian and headed onwards to Shanghai, pausing only to get lost on a precipice, climb the Hua Shan holy mountain and entertain ourselves with a spot of karaoke. We reached the unassuming town of Luoyang and navigated its streets. A few shops blinked neon, some had fruit outside, but mostly the town was made up of impression-less buildings. Not too far away, though, was something that certainly packed a punch. Inspired by the constant advertisements for 'Kung Fu Panda', we had scheduled a day off to visit the Temple of Shaolin where young shaven headed monks learn combat and gain unfeasible agility; practising the Iron Shirt, Crouching Tiger and Diamond Finger.

One day was to turn into three.

Overnight the storm had finally caught us and a thick blanket of snow had fallen across the city. Car crashes blocked the roadways, trees crushed road side stalls and men with heavy metal shovels made valiant attempts to counteract the pandemonium. China had been hit by the most severe weather conditions for 50 years.

Boll*cks!

As luck would have it, Charles had done a reconnaissance of the area and, with the roads under eight inches of snow, wandered off to make the most of a pre-located all inclusive breakfast. Enveloped in the blissful knowledge that we were unable to go anywhere fast, I curled back up in my cosy youth hostel bunk and dropped luxuriously back into heavenly sleep.

An hour and fifteen coffees later, Charles bounded back, a giddy cyclone of over excitement.

"Suse, you are going to love me!"

"Why? What? Why? What have you done?"

"It's more what I've found..."

We moved hotels.

For £10 a night we could stay in the youth hostel. For £12 we could stay in the place down the road with 18 course breakfast, sharks in the lobby, rose petals in the toilet and valet parking for our bikes.

Well, if we had to wait for the storm to pass....

Sadly, the storm did not pass as quickly as we hoped. After two days of DVD watching, enthusiasm for this new found luxury began to ebb into boredom, claustrophobia and a text-book case of cabin fever. More worrying than the tedium were the growing fears for our calendar.

We had flights out of Shanghai on the 24th November and a trip to SunTech, the largest solar panel manufacturer in the world, pencilled in for the 22nd. Having already missed out on two Chinese solar projects, I was adamant we weren't going to miss a third. So, despite attempts to keep ourselves occupied: "Jamie, is the yoga teacher actually humping my leg? It feels like it but I'm too scared to turn round!" we put down the remote control, re-corked the Sauvignon and made ready to depart from our opulent prison.

The snow was abating.

We would battle the ice.

15 November—Zhoukou, China
Day 185—15359.19km

Which is precisely how, having lost Charles in a snow engulfed village, Jamie and I ended up in a blizzard on the outskirts of Zhoukou.

It was only a few hours into the day's ride when Jamie got a flat. It was freezing. Literally. And so I waited with him and the gathering crowd of road workers while Charles, still a little slower, kept riding to keep warm. The tyre change took longer than usual. Numb fingers and tiny bits of shrapnel didn't make for easy work, so it was a while before we were back on the road, and a while later still that he had taken a different turn.

For Charles it was to be a blessing.

As that was when the storm hit.

Dark clouds hastened the onset of dusk. Muddy potholes filled with thin layers of ice which crushed as we rode over them, spraying us with frozen slush. Truck drivers, half blinded by the relentless barrage, swerved erratically, forcing us off the narrow road whilst first our brakes froze and then the gearing. We were stuck in a single speed, unable to go faster and unable to stop as we slipped and slided into the traffic. Snow settled on us as we moved. We were frozen to the core and in a constant state of perpetual danger. We passed nothing that would block us from the lashing blasts of ice and nowhere where we could stop and try to make a shelter. Vehicles loomed into our path through the darkness and we wrestled our bikes in a constant battle. It was hard to steer, hard to see and hard to know quite when the torment would be over.

I was nearly hysterical with laughter when we finally found a town and the incomparably welcome sight of an ever-fluorescent hotel sign. With no feeling in my feet or legs I swerved up to the entrance and wobbled across the lobby like a drunken maniac. Five minutes later we had been rescued by several curious and frenetic members of staff and packed off to a clean room, radiators blazing, hot shower at the ready. I put the kettle on while Jamie moved the splattered panniers into the bathroom where we could hose them down from mud and ice.

THUNK

"Oh God."

Power failure.

The blizzard had taken out the lines and the building was plunged into darkness.

"I'm sure it'll be fixed soon Suse..."

We spent the night shivering in any clothes that hadn't been drenched in the meteoric deluge. Teeth chattering and frozen to the core we slept fitfully through the night; skin like ice, mud caking our hair, chilblains itching like burning hot pokers.

The next morning we went to take stock of the damage. The road had not been covered by as much snow as we'd feared but the state of the bikes was a whole other matter. Still frozen from the night before, the lack of heat meant that we could do nothing to de-ice them. They were solid as rock and impossible to move. Without any power, we set about the task of chipping the blocks away with penknives, hammers and spoons, but it was to be four hours before we could hit the road and two hours more before I finally managed to get hold of Charles.

"Thank God you're ok. How was your night?"

"Great. I found a nice hotel and they let me bring my bike inside."

"It didn't freeze then?"

"There were garages all the way so I defrosted it with hot water."

"It would help me if you stopped talking now..."

19 November—Nanking, China
Day 189—15717.41km

The cold remained, but the further east we rode the more the snow and ice gave way to thick red mud and gentle rain. It was now clay, not ice, which stopped our wheels from turning, splattered into our eyes and covered us from head to toe. Scores of curious bystanders came to watch us in the villages we passed; Jamie valiantly navigating through the melee of China's back roads. I was grappled by the men and the guys adored by women. We drew crowds of up to 50 and were followed by eager school children. Florid faces peered out at us from sad, dejected buildings and motorbikes zoomed past us carrying cages packed with animals. Thick plastic curtains made for the doorway of restaurants and their soot-covered walls displayed old Chinese adverts. We stayed in grey forma kit hotels in grey forma kit towns, before setting off again into grey forma kit suburbs. There were welding shops, factories and huge concrete business centres all dotted intermittently with shining blasts of electric colour.

The signs of China's development were both a blessing and a tragedy. In the West, gigantic turbine blades had convoyed past our bicycles, all part of the fast-paced renewable energy sector ever-growing across the country. But in the East, coal showered us from thundering lorries and smoke spewed out over impoverished villages. Motorways cut through the centre of settlements and for weeks we saw nothing but the haze of pollution.

We froze in our saddles and cursed our way through our punctures. We ate the strangest things, the nicest things and bet each other to guess at our dinner's contents. We would slowly peel our socks off at night and in the morning peel them on again. We would stop outside shops, laugh with stall holders and build up the courage to make use of the toilets[33]. By the time we hit Nanking we were shattered and broken and the sub-zero temperatures had taken their toll.

With bad circulation, I had developed thick chilblains. My toes had turned purple and burst from my cycling sides. Swollen and splitting, two had deep lacerations and I would pray for the moment when the cold air would numb them. In the evening as we rested Jamie would offer to massage them— worried that I would never again regain the sensation. But the feeling would always return with a vengeance and then make me cry out with the exquisite itching.

And Jamie was in trouble too.

---

[33] Most are too distressing to describe but the most shocking was actually peeing on a rat.

The frozen air had settled on his lungs in the blizzard and he had wheezed and hacked his way through the smoggy miles. That night in Nanking he began to burn up, coughs ripping his throat and body wracked with convulsions. At 3am I got up and wrapped my duvet around me, hoping somehow to find a late night pharmacy.

The city was beautiful.

Nanking by night.

In all my gear and covered by the bed clothes I was warm for the first time I had been in weeks. Our hostel was by a river where boats were daubed with hanging lanterns, surreal fluorescent glow sticks framed the edge of the pagodas and the trees that lined the water's edge were scattered with twinkling fairy lights. I wandered on past and got directions from the A&E, eventually finding a chemist with its doors still open. As thin as a whippet and drowned under the duvet I then set about mimicking a man having spasms.

God only knows what the pharmacist gave me!

Jamie was marginally better the following morning. Pills swallowed and lozenges eaten, he pulled up his neck warmer and set off determined. Charles and I scurried along in his wake, he had the directions and the GPS was set. We had two more days to make it to Shanghai.

Susie's Diary—November 2009

Shanghai. Shanghai?!?!"

Somewhere back in the desolate sands of the western Chinese desert we stumbled across a magical truck stop selling delicious, steaming bowls of thick cheap noodles. Tucking into our lunch with a zealous gusto we started chatting to some truck drivers. As ever the conversation was limited largely to smiling, laughing and the words: 'English', 'Egg Fried Rice' and 'I don't understand' but with the help of some maps and gesticulation, we had finally explained that we were cycling across the country. The slightly inebriated truck driver rocked back on his chair and mulled over our intended destination.

"Ahh, Shanghai."

Before the penny dropped and his voice rose a number of octaves.

"SHANGHAI?!?! SHANGHAI?!?!"

Since then, SHANGHAI?!?!—always referred to in the same shocked, high pitch tones—has been our goal. Charles, who we found in a Homestay on the Krygyz/Chinese border, was planning to head to Beijing but not keen to cycle through even wilder snowstorms or miss out on our scintillating conversation, decided to join us on the journey south.

And—on Saturday night—after 16045.54km (9970.2 miles), blizzards, sandstorms, 50+ degree heat, -5 degree cold, 3700 metre mountain passes, 250km cycle days, numerous punctures, chain breaks, tyre explosions, stomach bugs, chest infections and the day spent getting stuck in mud, concrete and the kind of traffic only a city of 20 million can create, we made it.

21 November—Shanghai, China
Day 191—16045.54km

Two days before we would fly to America, we drew our bikes up to the edge of the Hangpu River and looked across at the Pudong waterfront.

Shanghai...

But there was to be no peace for those who get side-tracked by hotels with rose petals and shark filled fish tanks. We had one day to pack and the other to get a train to Wuxi for a tour of the SunTech photovoltaic (PV) factory.

SunTech, however, was a trip worth the sleep deprivation. Producing more than 13,000,000 solar panels for 80 countries around the world, it is the world's largest solar company. Though it exports 50% of its panels to Europe, a shift in subsidies and incentives back home are likely to see a much greater adoption of solar power across China, hopefully bringing about greater improvements in efficiency.

Even more excitingly for us though, was the fact that their Wuxi office hosts not only the biggest PV panel on a building— all 6900m^2 of it—but a basketball court, ping pong hall and pretty impressive climbing wall. We got a tour of the factory, saw the panels being made and watched amazed as an incoming shift were guided through a 'spirit boosting', 'mind enhancing', 'health and vitality' exercise class. I'd love to see someone try that at my work back home...

Back in Shanghai we could finally relax. From the sheer Tian Shan Mountains, across the 'Desert of No Return', onto the Tibetan Plateau and via the icy grey heartlands of industrialised China, we had finally reached the mind exploding illuminations of the country's most cosmopolitan metropolis.

We headed back to our youth hostel and back to find Charles. Through the long, sand filled nights and short snow covered days, Charles had ridden alongside us. He had buffeted the 'tent peg incident', joined forces with us as we dodged the tenacious hand of the authorities, and surpassed even Iain's record of food consumption at each and every sitting.

Though now unconcerned about cycling as a duo, it was still with sadness that we would leave our adopted SolarCycle team mate and jump ship for America. Really, now, there was only one sensible thing to do.

We were going out.

And I was ramming my feet in those high heels if it killed me!

Jamie's Diary—November 2009

"Put this in your fact pipe and smoke it!"

Timmy's quote actually dates back to our first week on the bikes in France. It seems an age away now, back in May, but relevant to my blog today.

I'm currently nursing a bit of a hangover from an obscure and rather heavy night out in Shanghai with some jazz musicians so there will not be too much chat, just some information for you to mull over.

- *Total distance cycled—16045.54km (9970.2 miles)*
- *Total time spent in the saddle—740 hours 12 minutes (equal to riding for one full month without stopping)*
- *Number of facts told to us by Iain—309*
- *Number of true facts told to us by Iain—9*
- *Longest day in the saddle—22nd October, 256km, 10 hours 12 minutes (Susie's less than popular suggestion)*
- *Average speed over the whole distance—21.4km/hour*
- *Number of crashes—7*
- *Number of crashes caused by Susie—6*
- *Number of days spent cycling—121 days*
- *Average distance cycled per day—131.1km*
- *Number of holes in Susie's best T-shirt—17 Fastest speed—72.53km/h (though Susie claims the her GPS readout at over 90km/h was correct, I'm dubious)*
- *Worst weather encountered—blizzard conditions in Eastern China approaching Zhoukou on 15th November*
- *Temperature of feet on the 15th November—approx..-5°*
- *Most precarious cycling conditions—riding on sheet ice by Deng Cheng 16 Nov (you could see your reflection!)*
- *Number of punctures for Susie—less than 5*
- *Number of punctures for Jamie—over 50*
- *Number of punctures for Iain—not sure but there was definitely one day when he got 8 in a row*
- *Number of Snickers eaten by Charles on a day off—5 (though I'm suspicious it may have been more)*

- *Hottest day—riding into Cairo 30th June, 45° in the shade (but the day I nearly melted in Tunisia felt hotter)*
- *Number of waves to and from stunned onlookers—3458*
- *Physically hardest day for Jamie's heart—6th June 09, 153.77km, 6hours 53 minutes of cycling with an average heart rate of 138 b/m*
- *Number of times Jamie has thought about a hot tub full of Hawaiian Tropic girls waiting at the end of the days ride—too many to count*
- *Number of years Susie has aged in 7 months cycling— 12 (the first waiter who guessed Susie's age said 24, (he lived) the last waiter said 36, (he didn't)*
- *Number of women dressed in Sexy Santa outfits it takes to rescue cyclists stuck in a lift—15*
- *Number of times Jamie moans when hungry / tired / hurting—constantly, apparently*
- *Number of accidents caused by motorists looking at Susie's skinny ass—3*
- *Number of laughs caused by Jamie wearing tights—300*
- *Most audacious wee stop—Susie in the central reservation of a motorway*
- *Site of the largest photovoltaic panels in the world— SunTech, Wuxi, Eastern China*
- *Number of days Jamie's been happy to get up and cycle—every day but 1*
- *Number of times Susie's NVQ in bike maintenance has come in handy—2*
- *Number of times this has nearly resulted in an injury—2*
- *Highest altitude reached—3705m*
- *Numbers of metres climbed in one day—approx. 3000*
- *Times Susie is amazed at how much food Jamie, Iain and Charles can consume—every meal time*

Western USA

Estimated Distance: 4213km
Actual Distance: 5873km
Sunlight Hours per Year West: Over 3000
Sunlight Hours per Year West: 2000 – 3000

Riding tunes:

"In the day we sweat it out in a runaway American dream
At night we ride through mansions of glory in suicide machines
Sprung from cages out on highway 9
Chrome wheeled, fuel injected and steppin' out over the line
Baby this town rips the bones from your back
It's a death trap, suicide rap, we gotta get out while were young
'cause tramps like us, baby we were born to run"

Born to Run, Bruce Springsteen

30 November—San Francisco, USA

"Mmmph—don't make me go."

"Come on Jamie, get up. We only have an hour to get to the ferry port. The woman from the council's coming. And a photographer..."

"But everything hurts!"

"Stop whingeing. I feel fine... technically that's because I'm still drunk—but anyway—we have to go!!"

It had been an amazing couple of days. We had landed in San Francisco and been promptly rescued by Birchy and Lulu from the solar company, Sungevity. They had cleaned us up, shown us the sights and provided us with a Thanksgiving feast dressed as Superman and Dolly Parton. We had cruised through broad leafy suburbs, past wooden houses and wide stoops, eaten oysters in beachside shacks to the north of the bay and looked out at Alcatraz and the Golden Gate Bridge. On our final night we had hit the Haight and a large bottle of tequila had hit us back. A friend who worked in a medical marijuana dispensary had given us a joint for our journey home and—well—when in Rome...

"Man, I can't believe we can cycle so fast. Sh*t this is incredible. I mean, I think we've actually left the city. All I can see is trees!"

Four and a half hours later, convinced we had pedalled faster than sound and half way to Seattle and back, we finally made it the 300 metres past the golf course. Two hours later still, the alarm had woken us from our comatose slumber.

I scraped my head off the mascara-smudged pillow and valiantly attempted to stand without falling over. We had an hour to get downtown, join a protest march, and head off for a photo shoot with Sungevity and some keen solar advocates from the Utilities Commission.

"Helpfully, I'm already wearing my clothes..."

For many years I had wanted to see the rolling hills of San Francisco. I'd seen cars fly down them, movie stars jog up them and trams speed out of control—all accompanied by the dubious sounds of tinny synthesisers and the visual sensation of six inch bouffant hairdos. My dreams, born of films of the 1980's, had not included seeing the rolling hills from a shaded spot in the gutter, thuds ricocheting through my temples while Jamie searched for a tent that had flown off his bicycle on a busy intersection; precisely where I found myself that afternoon.

With only two hours sleep and a curve ball of a day, it would be fair to say we were struggling with our first day's ride across the United States of America.

1 December—Redwoodcity, USA
Day 201—16121.78km

The next morning we woke up marginally more refreshed. Having pedalled late into the night we had finally given up on reaching San Jose and peeled off to one of the many motels lining the route. We sank some coffee and ate plastic wrapped doughnuts laid out in a small room overlooking the car park. It wasn't the most atmospheric or nutritious of breakfasts but it gave us the sugar boost we needed to pedal the final miles to the capital of Silicon Valley and G24, our solar sponsors.

Headquartered in California, the G24 team had been drumming up support for the trip and their CEO had invited us to visit. With particularly good timing, San Jose was hosting a nano-technology conference and so their PR guru, Ned, was in town and had rustled up support from some keen members of the cycling press. With more coffee and Ned's sharp Irish humour kick-starting our systems, we waxed lyrical about our adventures and the merits of our solar powered panniers before being bustled off for lunch and some much needed vitamins. Silently thankful that our cleated cycling shoes marked the reason for our dishevelled appearance, we greeted the CEO at a smart restaurant and settled in for a much longer meal than we technically had time for.

On the road four hours later, our inability to leave the vast platters of food inevitably wreaked its revenge, as day gave way to dusk to night. Naturally, the hills between San Jose and the coastal Highway 1 were bigger than we imagined, and obviously we had underestimated how long it would take to cross them. By the time we hit the final descent to the sea, it was jet black and we were kicking ourselves for doing precisely nothing about our near fatal lack of illumination— last inadequate on the icy roads of central China.

My front light was pitifully wan and Jamie's only marginally better.

I tried to ride alongside him so that combined we might be able to see but it was too dangerous on the narrow winding mountain roads. Instead I plummeted behind him, watching as the faint glow of his back wheel edged further into the abyss and ramming pothole after pothole into the chequered tarmac. Cars clearly shocked to find cyclists on a steep twisting path at night, veered wildly as they came across us and beeped their horns in shocked agitation. As they slammed their brakes on I gave chase, clinging desperately to their slip stream and the brief glimpses of hairpins now visible in their headlights.

It was a dangerous, exhilarating, death-defying blind descent. Like a 60mph computer game with obstacles flying at us from the shadows, huge chunks missing from the roadway and cars careering centimetres away as we plunged around the perilous switch-backs.

If I hadn't been so dehydrated, I might have weed my knickers!

Susie's Diary—November 2009

"What's Highway 1 like?" I innocently asked one of our G24 lunch companions, an avid cyclist and athlete.

"Well, there are a few undulations but it's a beautiful ride".

Undulations? Undulations? Are you kidding me?! Highway 1 would be more appropriately described as a 'litany of vertical inclines'. You practically need crampons to scale some of them!

Note to self: should you happen to be pedalling over 400 miles to LA, never take topographical advice from an Iron Man.

Highway 1, USA

Undulations aside, Highway 1 was nothing if not beautiful. Famous for its breath taking ocean panoramas, cliff grasping bridges and tall swaying pines it certainly lived up to expectations. We camped in the forest, basked in the sunshine and watched the cobalt blue surf crash into sheer rock walls. As the winding hills of the Big Sur finally gave way to the coastal plains we found quaint towns, blubbery elephant seals and handily appointed mock French patisseries.

We were back in the flow and churning out the miles. The sun was warm, the people enthusiastic—"Seriously, man, you have blown my mind!"—and, after happily stumbling across a carnival in Santa Barbara, had soon enough hit Venice Beach, with the neon lights of its Ferris Wheel to guide us.

We had been lucky in meeting Ned in San Jose. Though he worked for G24, he lived in Los Angeles and had insisted we stay when we arrived. And so it was that, four days since we left Silicon Valley for the Californian coast, we dodged rollerbladers, swerved miniature poodles and dragged our creaking wheels to Redondo Beach and up the hills to his home.

Susie's Diary—December 2009

Oh. My. God.

Picture the scene:

A log fire flickers sending a warm orange glow across the room. You are cocooned under a soft cashmere rug, enveloped in the sumptuous cushions of an oversized sofa. Two beautiful white Labradors nuzzle at your feet. A delicious smell of baking bread comes from the kitchen. To your right the news channel talks of 'snow, wind and driving rain'. To your left you look out across an incredible 180° panorama of Los Angeles Bay. A storm rages outside. Palm trees are bent double as the gale whips through them. The city is pounded by the incessant flow of water causing floods and mudslides from Malibu to Compton.

"Do you want to stay another day?"

7 December—Redondo Beach, USA
Day 206—16838.64km

Ned and his fabulous wife Marian had welcomed us in and immediately set about thoroughly spoiling us. If you need any more of a picture I should explain to you that I had the 'Princess Room' and required a foot stand to get into the four poster bed. We drank red wine, talked politics and listed of all the things that we missed back home—largely the sarcasm of classic British comedy. With children our age, Ned and Marian were worried about our meagre frames and began a 48 hour attempt to patch our wounds and fatten us up.

Whilst we had made it through the blizzard in China, I still had deep suppurating fissures in my purple toes and was only just beginning to pile back the pounds. Jamie's cough had abated in the clean coastal air but he too was dark eyed, undernourished and in need of a good rest. Mentally, too, Ned and Marian were a much needed tonic.

Though we had been battling the conditions and pounding out the miles, my personal goal to promote solar power was proving a much tougher and less quantifiable challenge. The vagaries of the political situation in Iran and China had limited our solar tracking success, problems with communications had lead us to miss two of our solar stops and the lack of common language meant my efforts to contact local publications in many countries had come to naught.

We had got off to such a good start; waved off by Boris outside London's flagship solar building, articles in the national press, photographed with the Mayor of Genlis at Solar Euromed and videoed at Petra to promote the 'We Support Solar' campaign's address to the House of Commons. Since then, though, failed opportunities, technical difficulties and linguistic issues had seriously curtailed our success.

Refuelled by Ned and Marian's encouragement and support from the teams at G24 and Sungevity, I was back on the case with a vengeance. Access to internet had me up at 6am each morning researching and emailing anyone who might be interested in the expedition; I wrote to the press in America, press at home and contacts we had made before we left. We updated our progress on our blog and on various environmental websites before messaging US solar organisations, providing articles to everyone and anyone we could think of and finally finding chance to talk to the guys at the University of Florida (UFL).

The libraries team at the UFL libraries had been watching our progress and months previously invited us to visit when we passed through town. The visit had grown in its ambitions and was now to be held in late January, prompting a week of solar and cycling events in and around the Gainesville campus. We had also heard from others in the solar and sustainability sector, now profiling our journey and the energy potential of the sun-baked deserts.

Two days later the skies cleared and, restored in body, mind and enthusiasm, we rolled out.

It will come as no surprise to find that we had once again given ourselves a stupidly small amount of time to cross a comparatively large expanse of land.

17 days and 2300km away lay Vanessa, Austin and Christmas.

The Wild Wild West, USA

"You want some more coffee honey?"

The waitress, wearing an apron, chewing gum and having painted her face on with the assistance of a trowel, stood above us brandishing tepid brown liquid in an spherical filter pot. Amazing. America had been beamed into my living room daily since childhood, but I had never expected the stereotypes to be so accurate. Turning east over the Laguna mountains, the landscape had slowly opened out and Baywatch beaches had given way to diners, road trains, and huge roadside casinos, signposts flashing above empty car parks. We were in frontier territory. Saloon door territory. People wore Stetsons. Without being ironic. It was like being on a constantly changing rolling film set, where the highway patrol officer warns you that "there's a storm a'comin".

We passed the train to Yuma, drank whiskey in Tuscon and rode late into the night accompanied by trucks so huge they might transform at any given moment. We clocked up the miles, fuelled by gigantic piles of pancakes, and drank cartons of milkshake so huge lifting them was a work-out. We passed sand dunes and cactus fields and faded metal road-signs; through tumbleweed and ranches, and by old wooden houses. There were porches and rocking chairs and rusty creaking hinges and lorry-size bill boards that were calling us to Jesus. Or at least:

Chad McLoud
Father
Lawyer
Warrior

Gila Bend, USA

It was the desert again, but a whole different desert. The Sahara had been scorching, vital and dangerous, the Sinai relentless, desperate and bleak and the Taklamakan, endless and clouded by haze. Here the skies went on forever and the colours were piercing. Every few miles there was a settlement or garage and the horizon was lined with electricity pylons and telephone masts. Signs of life were everywhere and we were back in our cultural comfort zone. For solar power, though, this culture and comfort wasn't necessarily making things easier.

Like it or not, the lack of democracy in China can get things done. If the powers that be want a solar plant, then they lay down the law and by December they've got one. Wind farm in a year? Just give me six months. In America, like Britain, it is not so easy to dictate what happens. Democracy brings with it rights and freedoms, but also a whole web and tangle of barriers. We might lobby hard for a wind farm or a nuclear power plant but then—erm—sort of hope that one doesn't spring up next door.

And protest if it does.

Solar in the US seemed a curious minefield. Imagine those amazing scenes from Thelma and Louise where they charge through the country in a blue Ford Thunderbird, dust clouds billowing along desolate farm tracks and straight past the monoliths of Monument Valley. Not quite the same if you add in a solar station. We were getting an inordinate amount of support, but the news and the politics were not nearly so simple. The people liked solar but they didn't always want it in their own back yard.

Along with Mexican runaways and proper ol' cowboys, however, frontier territory is nevertheless prime solar country and some of the towns have cashed in on that fact; Gila Bend, Arizona, being a perfect example.

With over 300 days of sun a year, the council announced Gila Bend a Solar Zone and pitched its support to build three solar stations. Looking past the changes this would bring to the landscape, they realized the importance it would have for job creation, and set in motion a construction boom bringing more work to the area than its local people could accommodate. Much like the gold rush of the 1900's, tradesmen descended on the region bringing a much needed boost to the economy, improvements to infrastructure and—in my mind at least—a clear increase in the potential for gun fights, leather chaps and a proliferation of Can-Can girls.

The Sunbelt of the US has serious solar power potential and if the trend is set then more stations will be built which can be used to sell energy across the states and into neighbouring countries for years to come. It might be an acquired taste, it might not look good in the movies but for America it could be environmental and business gold.

"The whole of the USA and populated parts of Canada may be powered from the south western states of the USA." – Desertec

16 December—Lordsberg, USA
Day 216—17849.87km

There was sun, there was warmth, but what we were worried about was the wind. With constant access to the internet, we checked the weather forecast daily, desperately trying to assess whether or not we had time to stop and rest our protesting thighs and bum cheeks. At Lordsberg we calculated a half day reprieve, checked into a motel and had a power nap. It was a Saturday so we figured we'd best have a beer too. We wandered down the dejected streets of the empty town in search of a suitable establishment before heading back to the motel's reception desk.

"Well Ma'am, there is one bar nearby. But you gotta get back to the motorway, cross the train tracks and head on down the end of the lane. It's 'bout three miles. But be careful y'all. Sometimes guests get lost. And they don' make it back..."

Less than reassured, we pedalled off into the night, finally reaching a square concrete building with a lone neon Corona sign. Straw on the floor and pool table in the corner, Jamie was the only man without a Stetson and I was the only woman under a size 18. We watched as the young guys from the local town vied with truckers for the attentions of two ladies who were certainly no strangers to the sight of a KFC bargain bucket.

It was an entertaining couple of hours where we chatted to the bartender, sank a couple of bottles and spent a prolonged period of time discussing how lucky we were not to live in the middle of the Texan desert. Partners in crime, we wobbled home laughing, relieved to have a rest from the punishing schedule and merrily drunk in the middle of nowhere. Together we had made it through highs and lows, thick and thin, faced every danger, solved every problem and rolled mile after mile after mile... and the following morning we continued to demonstrate an astounding capacity to annoy the sh*t out of each other!

"It's this way, Suse!"

"No it's not. It's this way!"

"You go that way then."

"I will. You coming?"

"No. I'm going this way."

"Fine."

"Fine."

I wanted to cycle one way through town, Jamie wanted to cycle another. He rode off his way, so I rode off mine. Sure that my route had been quicker, I stopped and waited for him at the edge of the highway. Certain that his was faster, Jamie did the same. Except we were at different intersections and there was no phone reception. Ten minutes later and we were both thinking the same thing: 'Do I stay here and wait, or do I try to find the other person.' Naturally, we both moved, now to different, yet still opposing, locations. This continued for about an hour. Until I eventually stopped and Jamie eventually tracked me down. We spent the rest of the day moodily cursing one another and being generally petulant. It was no-one's fault, but nevertheless the bickering set the mood for our one and only full blown argument.

Which went a little bit like this:

One and only full blown argument

Up ahead a tractor was stopped in the road. Its load of cabbages had fallen and the farmer was picking them up and putting them back in his trailer. It was a bright, sunny, early morning, easy pedalling through miles upon miles of flat grid-planned farm land. We were making good progress but our schedule was tight.

"Hey, let's stop and help this guy." Me: kind, caring, sensitive.

"We don't really have time. It's still 80 miles until Fort Stockton." Jamie: pragmatic, logical, sensible.

"But it won't take long if there are three of us."

"It'll take ages! And anyway, it's his fault for packing them badly. It's dangerous!"

"Well, if it'd take us ages, think how long it will take him."

"You stop. I'm not stopping."

"But if I stop it makes no difference as you'll have to wait for me."

"I'm not waiting for you. It's a long straight road. You can't get lost. Stop if you want."

"I'm gonna stop."

"Fine." Sulky silence.

"I thought you were stopping Suse."

"I can't. I'll lose you again."

Continue cycling. Fume silently. Mentally assess the chance of doing serious injury to someone if you pedal straight into the back of them at speed.

An hour later we took a break and the built up fury came tumbling out. Slights were made and words were pelted. I was upset that we had received an abundance of help throughout our journey but couldn't stop for one man in need. Jamie pointed out that the man in question had smiled, tipped his cap and was hardly in immediate danger, idly picking up cabbages in the gentle sunshine. I expressed—colourfully—that this was not the point. He maintained that, if we stopped for everyone I had wanted we would still be somewhere in Turkmenistan.

Then, to my distinct recollection, I drank a chocolate milkshake.

Seven and a half months and this was our first actual fight.

If only it could have been something more dramatic than cabbages!

22 December—20km from Junction, USA

Day 253—18935.46km

Though the road conditions in the US were certainly a lot smoother than other parts of the world, there was one issue we encountered continuously: debris. At some point the volume of fast moving traffic must have resulted in a number of collisions and the loss of various automotive body parts, wing mirrors and wheels. This endless flotsam led to a constant rash of punctures, largely for Jamie.

Ridiculously irritated by the continual stop starting, invariably on the hard shoulder of a busy, smoggy intersection, he had sought a solution. And that solution was Slime Tube Protectors[34]. According to the Slime website: *"The Slime Tube Protector is a lightweight, durable strip of extruded polyurethane that fits between the tire and the tube; protecting from thorns, nails and glass."* Which, to be fair to Slime, it ably did. Only as a slightly problematic side effect it also began to rub away the inside of the tyre causing a massive lesion and making it bulge like a snake that got over friendly with a hamster.

Naturally this became apparent about 40km from the town of Junction, as the sun began to set in the late afternoon...

Susie's Diary—December 2009

We ate doughnuts and debated our options.

I could cycle on and try to get a new tyre, Jamie could hitch back to the previous town to nab one from a bike he had spied by the roadside or we could attempt to fix it.

Back in Uzbekistan our fantastic and knowledgeable host, Hans Bedowski, told us of the trouble he had while cycling in West

---

[34] I forgot the name of the tape while writing this... it was not without some trepidation that I typed the words 'slime, rim and tube' into the Google search engine.

Africa. His bike would pick up punctures from Acacia trees, rusty nails and bad road surfaces almost daily. Those of the locals would not. One day he asked a friend how he avoided punctures. He was told that he simply needed to line his tyre with goat skin.

Back on the Texan roadside we recalled the tale.

We looked around.

To our left was a deer.

A dead deer...

And right next to the deer a handily sized, tough yet flexible piece of rubber.

Using the remnants of our Iranian super glue, Jamie stuck it to the inside of his tyre and 5k later, as the pressure began to build and the rip widen, we found a new bit of rubber and patched it again. And again.

We continued well into the dark, desperately scavenging for suitably flexible detritus in the lights of the lorries thundering inches from our foraging fingers. Eventually we admitted defeat and set up camp.

23 December—Friedricksberg, USA
Day 254—19062.19km

We struggled up the next morning and cruised the last few miles to Junction at a pace that Jamie was happy to fly off at. Thankfully there was a hardware store and, while I surreptitiously nicked his wet wipes to cleanse my clammy armpits, Jamie found a replacement tyre and secured it in place. Snacks eaten, relief noted, we headed out of town across a wide metal bridge and pedalled straight for the Texas Hill Country.

Not that we knew it was Texas Hill Country at the time of course, but this was something that became immediately apparent due to the road signs claiming it to be 'Texas Hill Country' and the profusion of hills which accompanied them. It slowly dawned on us that, with the delay and the late start, we now had just over 24 hours to ride 250km to Austin through Lance Armstrong's training playground.

Interesting.

An old friend, Vanessa, had arranged for us to spend the festive season with her folks in San Antonio and had left her car keys at her apartment for us to pick up the next day. We needed to reach her flat in Austin by four and drive the 150km to her sister's house for six o'clock the following evening.

With the vigour and determination that only the thought of an 'all-the-trimmings' roast dinner can induce, we pounded up and down the undulating tarmac and—pausing only to take photos of my festively graffitied '17 hole' T-shirt for the SolarCycle website—reached Fredericksberg by nine, ordered a pizza and bolted ourselves in a motel for the night.

Susie's Diary—December 2009

On Christmas Eve we had 80 miles to go. We had been pedalling 100 mile day after 100 mile day, battling the wind and trying to eek nutrition from burgers and energy drinks. The days had been plagued with punctures even before Jamie's tyre had ripped from the inside and we were forced to fumble around the roadside to patch it up with debris from exploded vehicles. We were in the area affectionately known as 'hill country'. A storm was coming...

Dragging our exhausted bodies out of bed, we trudged through the rain to the reception for a sugar and caffeine boost. As dawn broke, the dark gave way to ominously grey skies and revealed the palm trees lashing from side to side. Donning our warmest gear, we got ready to leave.

Flat tyre.

Our 7am departure time slipped away as I wrestled to remove and replace firstly the layers of arctic gear and then the inner tube. All set, we made to leave once more.

"What's that noise Jamie?'

"What noise?"

"Is it the wind?"

"Oh that. Yeah, it's definitely the wind... the wind and your other tyre..."

After months of being the one who never got flats, I now had two in a row. This was not my morning. Half an hour having slipped past, we made it out into the cold and hit the road. Almost literally. The wind was blowing at 40mph as a constant and 50mph in gusts. With ridiculously good fortune, though, in our direction.

Wooo hooo!

We were flying.

The rain had tempered and become occasional spots. The road was good and either not that hilly or we were being hurled so fast we hardly noticed.

Cajunk. Cajunk. Cajunk.

Puncture 3.

8.37am

Freezing and having trouble standing up straight in the howling blasts we wrestled to change it in record time. The clock was ticking... Luckily for us, though, other than a bit of hail and me getting catapulted off the road on a corner, everything else went to plan.

In fact, against all odds, we made it by 3.30pm and had time to shower, change and have our first road trip. Well, first road trip where we weren't self-propelled at least. And we made it bang on time for Christmas Eve complete with tree, presents, eye watering rum cake and readymade family!

Austin, USA—Christmas and New Year
19189.44km

The next few days were a blur of food, drink and kindness. We were welcomed into Vanessa's family with open arms, and by her friends with open bottles of Jack Daniels. After a beautifully relaxing Christmas in San Antonio and a trip to the museum at the Alamo, we drove back to her house and our kit and spent the New Year in Austin, being tourists, partying through the New Year and generally causing mischief.

We had our bikes fixed by the guys at the Bicycle Sports Shop, danced until dawn with a naked man dressed as an angel, talked at a meeting for a group cycling bicycles to Mexico, sat in the deep leather seats of the state capitol building, watched our friend play guitar in a gig on 6th Street and spent long merry evenings in fantastic company.

And—for once—there was a spot of romance.

On arrival in Austin, Vanessa had quickly assessed that Jamie and I were not an item, and promptly listed all her single friends. It transpired that, of those we would meet, there was only one unattached young gentlemen; fortunately for me, a tall and dashing one. Having been figuratively—and literally—in the desert for some time, he was duly wooed and promptly kissed. Or, as Jamie would put it, "you grabbed him and dragged him off like a cave woman!" On Jamie's part, the morning of our departure he arrived back at Vanessa's at midday, clothes on from the night before, big grin ingrained across his face.

"Do we have to leave today? I haven't had much sleep!"

Like I said, romance...

Susie's Diary—January 2010

It was pretty difficult to leave Austin after such a great few days (particularly yesterday when Jamie refused to), so thank you so much to all of the fantastic people we got to bring the New Year in with for supplying a plenitude of margaritas, karaoke and an awe inspiring take on the Dirty Dancing lift.

After a manic few days at Christmas it is, though, amazingly peaceful to be pedalling again. Daydreaming, watching the world go by and hanging out in the... sunshine.

Finally!

First thing's first though, Houston.

Not actually somewhere we had planned to go. Naturally. But at 7pm last Wednesday, in the dark, as the rain began to fall and the temperatures plummet we rocked up somewhere in the northern suburbs. Attempting to read the GPS in the blurry half-light a concerned homeless lady wandered over.

"You guys lost?"

"Erm, are we? Yes."

"Well, do me a favour and don't go any further. This motel isn't too bad. They rent by the hour but at least the crack dealers won't bother you."

We stopped.

It was everything you could imagine of a seedy motel but it was warm, dry and—amongst 'other' things—had Aladdin on TV. It was my birthday so Jamie braved the weather to get a cake and some food.

Lucky, really, as I was too scared to leave the room.

We awoke the next day to the 'arctic blast' that had hit the South. Since we had a load of time to get to New Orleans we decided to hole up and stay put. The decision was made much easier by the fact that a friend from home had just moved to Houston for work and to terrorise its Southern belles. He would rescue us that evening.

Thank God!

And so it came to pass that—just a short while after our less than salubrious introduction to the city—we found ourselves in a loft apartment with a hot tub on the roof overlooking the sky line.

Very Miami Vice.

If Miami was in Texas and didn't have as much vice in it.

9 January—Houston, USA
Day 240—19463.13km

After a night with our hosts in Houston, they all decamped to the ranch for the weekend (seriously, they had ranches), leaving us to our own devices.

Leaving us to our own devices, that is, in the apartment with stainless steel kitchen, floor to ceiling windows, deep comfy sofa, power shower, 408 TV channels and bird's eye view of the city... Still reluctant to get back on the frozen highway, and as we had acres of time to get to our next solar project, we passed the day being cultured in the museum district before finding a bar and becoming acquainted with its cocktail menu.

By industrial strength margarita two, we were in a heart to heart discussion about life, love and all things related. Jamie had been in touch with his lady from Austin and—it being a Saturday—she was driving down to join us.

"That's cool right?"

"Yeah—totally..."

The apartment was open plan with the bedroom a mezzanine above the couch below.

Jamie decamped to the hot tub, climbing through a window and presumably giving the CCTV and security guard an eyeful.

I settled in for a night on the bathroom floor; the only room that had a door on it!

Highway 10, USA

I wouldn't call Highway 10 the most relaxing ride of all time but avoiding it would have meant a 300km detour. Hardened to the onslaught of traffic and used to the motorways of China, we perused the map, peered down the road ahead and decided to take our chances. What we hadn't realised from that safe, calm highway after Houston, though, was that when you reach the bridges over the deep swamps of Louisiana, the challenge is somewhat exacerbated by an absence of any hard shoulder.

In case you are ever considering it, I can confirm that there is definitely something exhilarating about riding along the narrow precarious roadway which rises above the murky waters near Baton Rouge—namely, your astounding proximity to death.

Imagine the biggest American lorry you can. Now imagine another, a metre from it. Both are traveling at speeds exceeding 70mph. You are cycling alongside these Goliaths of automobile engineering with absolutely nothing but a jumper and a '17 hole' festively decorated T-shirt to protect you. As they shudder past your ears they are immediately replaced with others, just as fast and just as huge. You are on a never ending grey concrete bridge. There is nothing between you and the wall which stops you from tipping off into the depths below. There is nowhere to stop. You cannot turn around. Shards of metal ping from underneath monster wheels, trees rise like zombies from the dirty swamps and fallen branches litter your path in a treacherous slalom of un-navigable debris. You are buffeted by the powerful vortex surging from the gigantic machines just centimetres from your shoulder blades. It begins to rain. The wet road loses traction as you lose visibility. Cold spray flies into your face as your ragged clothes cling to your back in the half light. There is fear. And panic. And nothing you can do. You have no idea when it will end.

Thank God for headphones.

I mentally swept across the situation, vaguely dwelt on our chances of survival and promptly turned my music up, transported myself to Hawaii and sang AC/DC tracks all the way to town.

Well, nearly all the way to town.

Five miles from Baton Rouge, Highway Patrol finally caught us.

We had stopped. A four lane, almost vertical road bridge lay between us and the city. To join from the angle we were coming at would mean entering the traffic directly between the trucks careering along Highway 10 and those swinging east from a southern road, all of which would be traveling at least 50mph faster than we were. We would then need to either stay in the very middle of the traffic or cross two lanes of it to get to the right hand side, where there was around 30cm of gap between the road markings and the concrete wall.

"We can definitely make it."

"Suse, I'm thinking no."

"We've never not made it before...."

Look of general exasperation.

"Ok. Ok. But we don't have any other option. We get across it or we ride 100km backwards down the swamp road of death."

Highway Patrol were a welcome sight.

13 January—Baton Rouge, USA
Day 244—19938.8km

Tumbling out of a police van was probably not the way that Jeff from the Louisiana Solar Energy Society (LSES) had envisaged us arriving, but he took to the situation with aplomb. I had emailed him a few days before and asked if we could come for a visit. He had contacted the local press to meet us, offered us a roof for the night and happily informed us that Wednesday night was pint night.

Perfect.

CEO of Gulf South Solar, Jeff had begun working in photovoltaics after hurricanes knocked out the grid power which serviced his home. His enthusiasm for solar as an independent power source saw him establish LSES and petition Local Government to enact legislation allowing solar electricity producers to sell excess power back to the national network; much as the 'We Support Solar' campaign was aiming to do in the UK. It was great to hear tales of his perseverance and success, as well as those of his work with Brad Pitt installing solar panels at the New Orleans, 'Make It Right' Foundation.

"Wow—I can't believe you got the state government to change their policy. So, erm, does Brad happen to be in town at the minute..."

Bleary eyed, we woke up the next day to the increasingly familiar sensation of a different bed and a mild hangover. Jeff insisted that we could not leave Baton Rouge without taking a tour of the University and—welcoming any excuse to shirk some mileage until we regained full mental facility—we jumped at the chance. Two hours later we wobbled our way out of town pondering strange tales of sororities, college football and the best ways to feed a 600lb Bengal Tiger.

14 January—New Orleans, USA
Day 248—20099.58km

Since Austin there had been a complete 180 degree shift in our approach to cycling. Having trained ourselves to ride long distances on meagre fare, now we could not want for more on our towering plates and had days to spare before our next stop at the University of Florida. In New Orleans, one of Jeff's colleagues had put us in touch with her relatives, Maite and Lee, who had offered us a place to stay and encouraged us to explore the city. Its 25 cent Martinis included. We accepted with relish and spent an astounding weekend wandering between Creole townhouses, listening to soul singers on street corners and embracing the French influence in its entirety by diligently spiking our own drinks with absinthe.

Dive bars and blues bands aside, we tried to maintain the educational nature of our journey with a visit to the Make it Right Foundation homes, for which Jeff (and Brad) had installed the solar panels. The foundation, established to rebuild the Lower 9th Ward after Hurricane Katrina, was a fascinating insight to the terrible devastation the hurricane had wrought and a stark reminder of nature's true power. When the levees had broken, vast swathes of the city had been literally washed away, the 9th quarter amongst them. Jeff had recounted a tale of the Baton Rouge football stadium doubling as a refugee camp. Maite told others of lives lost and families torn apart. When we finally dragged ourselves away from the music and the madness of the city, the remnants of homes, cars and boats stilled lined the road east, smashed to pieces by the hurricane, hundreds of metres from the shoreline. Four years on, every house still standing was up for sale, most still bearing the marks of the ravaging wind and battering seas.

I was in an unusually sombre and contemplative mood as we pedalled past reminders of the destruction on the way to Biloxi. In small part, though, as this was also the second time in eight months that I had felt unwell.

Susie's Diary—January 2010

We stopped at Fort Pike for lunch. I couldn't eat. My stomach was in my throat. Back on the road to Biloxi, progress was not much better. We stopped for me to lie down, and to check the map to see what was ahead. Nothing for 30 miles. Hmmm. Three miles of excruciatingly slow pedalling later we stopped again. I had no energy. I lay on the roadside in the foetal position as waves of nausea swept over me.

"Leave me. I'll probably die. Save yourself..."

I laboured the point for effect.
Between the stabbing pains I took a sip of lemonade. It was all it took.

BLEURGH!!

Disgusting.

We were still 27 miles from Bay St Louis but feeling marginally better with nothing in my stomach, we plodded on.

19 January—Bay St Louis, USA
Day 250—20222.71km

I crashed out in the nearest motel while Jamie went in search of supplies and a gigantic bag of Jelly Babies. Overnight, another storm hit. The rain was torrential, lashing down with an incredible force and providing a tiny insight into the extreme weather conditions that pummel the coastline of the southern USA.

The next morning I peered out of the motel room door into the deluge and wondered, now feeling fine, how far I could push the sickness excuse. We hung out for a while before Jamie took decisive action and began to put on his wet weather gear. Jeff had put us in touch with his brother, Kevin, who lived in Mobile, and Kevin had generously offered to give us food and shelter for the night. We were due to reach Mobile that evening and to speak at his son's school the following morning.

As per usual, things did not go to plan.

The rains got stronger and stronger.

The water went straight through our rain jackets, straight through our panniers, straight through the wiring on our solar panel batteries... Our soaked skin turned white and wrinkled in the biblical downpour. The wind was driving against us; hurling water into any orifice we had the misfortune of leaving uncovered. We couldn't see the road ahead and the cars on it could not see us. We got stuck in sand drift after sand drift before Jamie finally got a puncture. I tried to ask if he needed help but my mouth filled with water before I could get any words out.

It had taken us four hours to travel 25 miles.

I took the executive decision to refuse to go any further.

20 January—Biloxi, USA

Day 251—20266.69km

Unfortunately, the next day it was Jamie's turn to feel unwell. He had been up all night with toothache and we were forced to make an emergency stop at the dentist's, most likely a direct result of the gallons of fizzy drinks he had consumed in the preceding months. Luckily though, we didn't have far to pedal and the rains had left a fresh, crisp day with clear blue skies. We stopped by a dock side and ambled through towns of blocky houses on block style streets. The ride was gently rolling and relaxed so, despite the dental detour, we were making good time when I got a flat. Jamie—exhausted from a restless night—lay out in the sunshine while I laconically set to work.

The first car stopped.

"Ma'am—are you ok?"

"Yes, fine, thank you."

Then the second.

"Everything ok there?"

"All great—thank you though."

Then a third.

This was getting strange.

Jamie, head nodding to the tunes playing on his phone, lay oblivious to the activity and the dangerous looks cast pointedly in his direction. I waved and smiled at the slowing motorists in an attempt to dilute the consternation and deflect the daggers winging their way towards his prostrate body. Eventually though, it was all too much for one guy who pulled over, slammed the door of his pickup and put his cap on as he stormed across.

"Ma'am—you have to let me do that for you."

I once again declined with thanks.

He was seething but I remained impassive.

"Well in that case Ma'am, I've just gotta ask one question."

"Of course."

"Why isn't HE doing this?!"

Jamie was gyrating on the ground, eyes closed, muttering along to the Beastie Boys. The chivalrous Southern gent glared at him with an unhinged mixture of venom, intent and utter disdain. I toyed with several responses but in the end settled for saving his skin and happily amusing myself for good measure. Leaning in and adopting a conspiratorial whisper I confided:

"Well—you see—he's a bit special...."

Susie's Diary—January 2010

On the last leg in to Mobile, Kevin called. He had been in touch with the local news station who would meet us on the road to do a story. Half an hour later and we spied a camera on the verge. It was the man from Fox 10. We videoed a short interview and chatted to Stephen, a guy passing who was interested in our journey. As we pushed off he called out to Jamie.

"Hey man, I don't know how safe it is cycling round here. Do you want to borrow a gun?"

Such a kind, yet such a truly strange offer. Jamie responded as only an Englishman would:

"Erm. No. But thank you very much. You know, I would love to but—um—we don't have much space in our bags..."

A few minutes later we pulled into town and followed the GPS to Kevin's house. What a stunning home; a huge white colonial townhouse, with wooden porch and wide arching entrance hall. We were welcomed in and had a fantastic evening with his family, laughing, drinking and eating fresh Mexican food. Right up until the time for our news debut.

We decamped to the living room.

Fox 10—Nightly News

9.15pm *Not up yet but there was a good feature on the dress code in a local school.*

9.29pm *Still no sign of us, but there had been a big car crash.*

9.30pm *The local council made an important ruling on pavement levels, obviously that would supersede.*

9.52pm *Well, there was bad weather and people need to know about the weather.*

9.58pm *"And after the break, our last piece on some very adventurous people". (Finally our moment of fame and fortune was imminent...)*

9.59.5pm *"Local men in Beijing brave the icy temperatures to take a dip in the river."*

What?

What?

Dip in the river?!

They weren't even in Mobile.

They were in Beijing. Going for a swim!

We were cheated.

Totally cheated.

22 January—Mobile, USA

The next morning we went to school with Stephen and Walker, Kevin's two sons. We had been asked to speak about the trip in their morning video broadcast which was, naturally, set in a dog kennel. (Whatever happened to assembly?) We happily obliged despite the fact that we had absolutely no idea what was going on and the only people that did were a gaggle of enthusiastic under 12's. After a nicely surreal start to the day, and a massive omelette, we bid farewell to our generous host and pedalled off into the sunshine. Finally hitting Florida a few hours later we once again smelt the salt of the ocean and headed for the beach front in the holiday town of Pensacola. We cruised along the main street and looked up at the flashing neon hoardings beckoning us in to try their beer or eat their seafood chowder.

"Jamie—seriously—do we have to go to Hooters?"

"It's the cheapest place in town..."

Having seen the alluring orange letters, my team-mates eyes and thoughts had wandered and he made a valiant and enthusiastic argument for the financial merits the establishment had to offer. After forcing him to assess other, more salubrious, options I had to concede that it was indeed the easiest on the wallet and pulled up a seat up at the bar. Partners in crime we then set about taking stupid pictures through the bottom of our beer glasses. We were on the home straight, in our final state at the end of our last country.

Though we'd had a blistering ride across the Western states, in our days off since Christmas we had eaten Creole in Baton Rouge, watched rag time bands under the wrought iron lattices of New Orleans balconies, drunk neat liquor in seedy basement bars, and spent long evenings in the type of flats you see in *Californication*. Now able to ride 100 miles without breaking a sweat and, for once, on track with our solar schedule, we admitted the guilty feeling of being less on an expedition than on a slightly energetic holiday.

Which meant that it was nothing but amusing the next morning when we set out in the face of the most ridiculous adversity.

In case of any doubt, the best place to be in the face of a raging storm, bearing down on you like the God of Thunder, is definitely not a narrow sand bank with nothing between you and the tempest hurling itself across the Gulf of Mexico.

On Santa Rosa Island the heavens opened up again, battering us mercilessly and forcing us to take refuge in Panama City.

"The rain's going to stop the day after tomorrow. I think we should wait it out."

"Is this because you've seen another Hooters?"

25 January—Panama City Beach, USA
Day 256—20644.83km

Waiting out the storm meant 130 mile days back to back, but with Lady Luck on our side we awoke two days later to the conditions we'd been dreaming of: sunshine, following winds and stunning, flat, smooth, coastal, tarmacked roads.

If you are a cyclist, you will remember days like these, days where you get totally lost in the movement. No howling gales or terrifying vehicles to pummel you, no swerving exploded car tyres or dodging potholes as you mentally write your own obituary[35]. Just you, an empty mind and the road rolling past like a never ending Super 8.

The journey from Panama City was text book, and involved picnics, old wooden jetties and petrified trees, rising eerily

---

[35] Died at 90 in a freak base jumping accident and is currently being mourned by her attractive grandchildren. One of whom has a Nobel prize.

from the ocean floor. We passed an air base of jet fighters flying overhead, ate huge seafood platters in small town restaurants, camped in the forest and stopped by the water's edge before—in an unprecedented and never to be repeated move—hitting a deadline without drama on the 27th of January.

That morning we wandered into a rock and roll diner, slung ourselves into deep leather booth and ate a giant stack of blueberry pancakes. After a leisurely breakfast we grabbed our panniers, jumped on our bikes and ambled the three miles to an intersection at the edge of the town of High Springs where, at 10am, we were escorted to the University of Florida by the day-glo thighs and happy chatter of the Gainsville cycling club.

27 January—University of Florida
Day 258—21092.57km

Early on in our journey, the library staff of Florida University had read about our expedition and asked if we might pedal past the campus on the last leg of our trip. Though we had readily agreed to try at the time, since our arrival in the US, we had been able to coordinate with them and become the key speakers in a solar power and cycling-based week of events, inspired by the expedition. We were honoured, delighted and, in my case, excited to do something other than send articles into the ether about the solar aims of the expedition.

Eight months and 21,000km since we had left London, we were deposited at the front of the University by our lycra clad convoy and handed over to the SolarCycle week organiser, Christine. Our hour-long presentation on concentrating solar power, cycling, SolarAid and exactly how stupid it is to detour onto the Tibetan Plateau, would take place the next day, giving us time to relax and take a tour of the buildings.

Amazing.

A real life American University.

I have to admit though, it's a strange feeling when you enter a room and see your own face looming down at you from three life size posters.

"Exactly what have you got us into this time, Suse?"

SolarCycle Week, University of Florida

Clearly, the PowerPoint I had cobbled together at Christmas did not work. Naturally the computer I was working on crashed. And obviously I did not help matters by spending the previous evening at a basketball game, but at 4am—and with the help of numerous pale faced late night occupants of the University computer room—I finally re-wrote our presentation and saved it to a stolen USB stick.

The talk itself passed in a blur. Probably a direct result of the four espressos downed in quick succession between 9am and 9.55am. But we discussed the potential of solar energy, babbled about frostbite and ravaging winds, and showed photos of head scarfs and six inch stilettos, before finding ourselves outside again, blinking in the morning sunlight. We were recorded, interviewed and invited for a spin on a solar bike before being duly thanked, given a packet of crisps and told to enjoy ourselves. It was a heady couple of hours.

Luckily, enjoying ourselves was not going to be a problem. We had met the world's most energetic, insane and hyperactive man on the way round the campus who had immediately signed up to become our tour guide. Filipe was full of the levels of enthusiasm only manageable at the tender age of 22, and insisted we could not leave without seeing what life was really like at an American University.

He had one thought in, mind: Beer Pong.

For anyone that does not know, Beer Pong is a drinking game where players take it in turns to throw a ping pong ball across a table into a cup of beer at the other end. If a ball lands in the cup, then the contents of that cup are consumed by the other team.

There are a lot of cups and—usually—a lot of games...

"It's 5am. I've gotta get home Jamie."

"Where's Felipe?"

"In the garden being sick behind a bin."

"Yeah, it's probably a good time to bail!"

Those young ones bounce back though. The next morning, destroyed, I spent a long time eating bread-based products while Jamie's eyes protruded like golf balls from the midst of his pained visage.

Felipe and his friend turned up to say goodbye dressed as a banana and a gorilla, riding a tandem.

30 January—Ocala, USA
Day 261—21162.45km

The morning's ride was miserable. The fantastic Christine had arranged for another cycle escort, but this time the pelting rain and the fact it was a Monday did little for the turn out. There was Alex, a solar architect, a guy from the local council and Dave, a serious cyclist, unimpressed by our haggard presentation and keen to test the mettle of two upstart wannabes.

Alex was keeping us company to the edge of campus, the genial councillor to the edge of town and Dave as far as he could bear to torture us. All I wanted to do was stop, wait out the downpour and drink tea. Without a pressured schedule I eyed each motel and restaurant we spluttered by with ill-disguised hope and longing. Sadly, Dave, head-down, determined to set a strong pace into the driving deluge, was a man on a mission. He rode with us past every opportunity to stop and deposited us in the middle of nowhere, feeling pitiful and looking much like we had been dunked head first in a swimming pool.

"Jamie, it would be fair to say this is a low point."

As soon as we could, we took our sodden possessions to the nearest Motor Inn, ordered a takeaway and spent the rest of the afternoon watching bad TV.

"What's wrong with Ping Pong? Why couldn't we just have played Ping Pong...?"

Susie's Diary—January 2010

"But no, Orlando. You just don't fit. I know we could snuggle and get on but it's just not practical for pedalling. What was that you say? You don't mind how impractical it is... But Orlando. It could be dangerous. Danger means nothing to you if you can be with me? Oh my. Marriage you say? And children... This is all so sudden."

Central Florida, USA

Unfortunately for all other visitors to the Orlando Tourist Office that day, it had a cardboard cut-out of the Hollywood superstar Mr Bloom, and was currently being occupied by two adults with a combined mental age of 14, who were using said cut-out as a prop for a series of ill-advised photographs. Ditto the next information stop, which had a child's pretend aeroplane. With such a short way to go, we had wound down to a point of extreme immaturity and were taking any opportunity to keep ourselves occupied and entertained.

Entertainment was, though, about to become a little more extreme.

Florida is a stunning and diverse state; from the crashing waves of the southern coast, the rolling hills of the northern interior and the beautiful orange trees that line the central roadways. It is also—if you decide to take a short cut through a farm clearly labelled extensively with 'No Entry' signs—home to a diverse range of wildlife.

Most alarmingly, alligators.

"If you get a flat, then I'm not waiting for you."

"Jamie! You'd leave me if I get a flat?"

"Yep."

"Just like that?"

"Just like that."

"What if you got a flat?"

"I'd still be fine."

"How come?"

"I'd run fast enough to catch you and nick your bike..."

Cheeky sod. Jamie was not helping my anxiety. I am not afraid of much. I'll climb mountains, run off buildings or hurl myself from a plane. But things that move quickly, in an unnerving manner, disguised as grass or planks of wood. Man alive, they make me jumpy.

2 February—Fort Lauderdale, USA
Day 263—21635.08km

At Christmas, Vanessa had mentioned that she had family in Florida and put us in touch with her sister's brother-in-law, Mark. We called Mark en route and were amazed to find that he had organised for us to dramatically ride our bikes into the Council Chamber that evening to receive a commendation from the Mayor. Naturally, though, we were now lost, on a tiny dirt track, surrounded by gigantic alligators and completely out of phone reception. We had not seen anyone for 20 miles and were using the unusual navigational measurement of 'let's go right as I reckon there's less chance of being eaten alive'.

Inevitably, a tropical storm arrived.

And so it was that, with four minutes to spare, soaked to the bone, scared out of our wits and with no idea what was going on, we wobbled our bikes straight into the Mayor's council meeting, narrowly avoided an elderly alderman and looked generally confused as we were commended, given a gilt embossed proclamation from the City of Fort Lauderdale and followed by a local TV crew.

21589km.

Never dull.

13 February—Sebring, USA
Day 275—21635km

'Woefully underprepared' is the only way to describe our first ever cycle race.

Though we had finally had success with the trip's solar message and ridden from London to the Miami the long way, there was still one more physical feat to contend with.

Back in China as we waited for the blizzard to pass, Jamie had found far too much time on his hands and located a mid-February 12 hour cycling endurance race in central Florida. Since we had been safely 13,000km away at the time, I had happily agreed to sign up and forgotten all about it. Now, with only a few minutes until the off, it was all too suddenly a stark and terrifying reality.

A grand total of four and a half hours sleep hadn't set us up as well as we'd hoped, nor had the preceding week spent by the beach. From Mark's in Fort Lauderdale it was 46km to Miami, and so the previous days had involved, surfing, sunbathing and celebrating with friends who had joined us at our final US destination. Other than those 46km and a few yards by California Cruiser[36], we had done little but eat, drink and convince the inhabitants of Del Ray to daub their faces in neon yellow face paint in celebration of the entirely fictitious 'William Wallace Day'.

Now we would need to push all of that to one side, drag back a faint vestige of fitness and throw ourselves into one final test. We would see what nine months of endurance riding would amount to, if we had learnt anything on the road, whether or not we could hack it in a real challenge.

It was 6am on the 13th February.

We had borrowed some road bikes and were at the Sebring race track.

We had 12 hours to ride as fast, and as far, as we could.

Jamie's Diary—February 2010

Fifteen minutes until the whistle blew: I'd only just parked the hire car and taken the bikes out. Now we had to put the wheels on, attach the electronic tags and get the sports tracker set up on my phone.

10 minutes to go: "Where's my phone? Bo#^%cks! It must have fallen out of my pocket at breakfast." I left Susie to head to the line where they had already begun giving instructions for the race.

---

[36] Cruiser bicycles combine balloon tires, single-speed drivetrains, and straightforward steel construction with – err – looking awesome.

5 minutes to go: The phone wasn't in the car. "Sh&t!!!" Too late to do anything else. I headed back to the start.

3 minutes to go: Found Susie. Her saddle was in the wrong position and I'd gone off with the multi-tool. Everyone else was set, primed and listening to vital final instructions on how not to get lost on the course. Susie was frantically adjusting the seat post on her bike and I was standing there in my baggy trousers with a T-shirt under my cycling top in a futile attempt at keeping the chill from reaching my bones. I realised I hadn't got my race ID number on.

2 minutes to go: I was getting my race number pinned to my back by a kind spectator. Susie was fiddling with the saddle. We still hadn't checked if our electronic tags worked.

30 seconds to go: Susie passed me back the multi-tool and we checked our tags worked so our mileage could be recorded. I'd like to say that our countless hours on the bikes had prepared us and that, at this stage, we looked calm and professional. Unfortunately I can't. My trousers were tucked into my socks and my exposed arms were goose bumped. Susie had her hoody on with one leg rolled up on her track suit bottoms. Neither of us had lights for the Sebring race track, still pitch black. We clearly looked as though we'd taken a wrong turn and somehow ended up on the start line.

Go!

Off we pedal with nothing to be done about the numerous things we have overlooked.

12 hours to the finish.

12 hours to tally up as many miles as we could.

We weren't sure who to cycle with in order to keep a good pace but the second group that formed seemed to suit us fairly well. Pace was fairly swift but having racing bikes and no panniers to haul meant we were keeping up. By the end of the three laps of the Sebring race track it was almost light. We headed out on the rolling hills of central Florida for the hundred mile loop that would lead us back to the start line.

25 miles in: Flat tyre. Well, not flat, but very low and after 25 miles I stopped to pump up it up, get food from my saddle bag and take a leak. I told Susie to stay with the pack so as not to lose energy. I intended to catch them. Not as easy as I'd thought. Conditions stayed cold and the wind was directly in our faces. After 20 miles of pushing and almost catching the pack I resigned to the fact I wouldn't manage before the turn around point and eased off slightly, reserving some energy.

9 hours to the finish: Susie was waiting for me at the turn and we took ten minutes to chomp some food down while I got my tyres properly inflated. Susie got her seat post adjusted again.

7 hours to the finish: My saddle broke and started to move from side to side. Nothing too worrying but it was irritating. We headed back to the race track and roughly the half-way point. This time we took it in turns in the lead as we had been doing for the past 9 months. The wind was with us and we kept a good pace.

6 hours to the finish: Lunch time. It wasn't a pretty sight. Both of us trying to get pasta, banana and sugary snacks in our mouths as swiftly as possible without it coming straight back out. I went to fill up my water bottles only to find on my return two of the mechanics fighting to massage Susie's legs with their 'special stick'. Susie finally got her seat in the right position and popped some painkillers for her knees, now killing from the previously cramped riding position.

Ten minutes rest in total, then straight back to the job at hand: racking up the mileage. My aim was to try and see if we could do over 200 miles in the allotted 12 hours. This meant a high average speed and precious little time to stop. The next part of the course changed, smaller circuits of 11.7 miles. The sun had emerged and we found the cycling pleasant. Our first 30 miles was great. Susie would invariably be singing out loud when riding behind me in my slipstream. In fact I was getting a little embarrassed as we passed rider after rider with her screaming out lyrics from behind, as if saying "this is a breeze, I'm passing you and not even out of breath".

Show off.

I took this as a cue to pick up the pace a bit as she obviously wasn't working hard enough. After the next and final break though Susie started to feel the burn and was flagging on the uphill section.

2 and a half hours to go: Over three quarters of the way through and I was expecting time to drag and my momentum to drop. However, now used to snacking on the go and pushing our pace, I was surprised to find I had lots of energy. In the last couple of laps I took the majority of the headwind.

40 minutes to go: In the final hour we re-joined the race track to complete circuits of 3.1 miles. We figured we needed 4 circuits to get us over the 200 miles. Just about possible. Unfortunately Susie's energy levels were low and she urged me to try for the four laps without her. I sped off like a madman, surprising myself how much more I had left in my legs. It wasn't enough though and I could only make three circuits in the time we had left. I then waited on the finish line hoping Susie would make it round 3 laps as well.

2 minutes to go: Susie crosses the line after an exhausting final lap.

Round the world

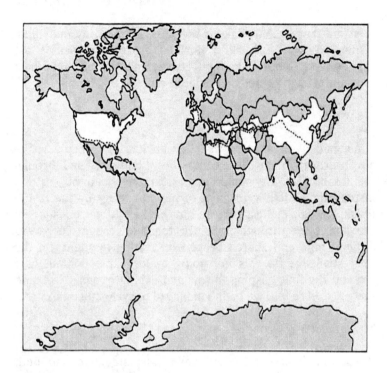

Estimated Distance: 19,680km
Actual Distance: 21,953km
Sunlight Hours per Year:

"Within 6 hours desert receive more energy from the sun than humankind consumes within a year"—Dr Gerhard Knies

Riding tunes:

"If you had one shot, one opportunity
To seize everything you ever wanted-One moment
Would you capture it or just let it slip?"

Lose Yourself, Eminem

19 February—England

Day 281—21918.89km

It is said that no snowflake is the same. Each a tiny individual spider web of exquisitely formed frozen arteries. I sat mesmerized. Gentle flake after gentle flake drifted down past the window and settled on the ledge outside.

We were home.

Jamie had been brilliant that day at Sebring. More a rogue adventurer than hard core competitor, I had struggled through the final miles. He could have steamed ahead but settled instead for pedalling in front of me and dragging me to the finish. It was a selfless gesture but one he had not questioned. We had been through so much together, fought for every difficult mile and shared every exhilarating moment. On the final challenge he was not going to leave me behind. As I crossed the line, gasping for air and aching in arms, legs and body, we were told we hadn't managed to cycle 200 miles.

The official was wrong.

When our final details had been calculated, we had both reached 204, each winning our categories.

Now back in the UK, in the heart of his family with mine in tow, Sebring racetrack was already a world away. We had less than 50km to ride to City Hall, alongside black cabs, red buses and curiously familiar road signs. We would be met by the team from SolarAid, Nokia and SolarCentury as well as friends, family and any press they could muster.

The night we touched down on British soil, I sat in Jamie's aunt's house writing an article for the Guardian while Jamie put the bikes back together. I was still doing everything I could to profile the expedition, while he was still doing everything he needed to make it happen.

We awoke to a soft blanket of snow gently resting on the road outside. Piling all our belongings back into our panniers we packed up our kit, set the GPS, and got ourselves ready one last time. Jamie's mum and aunt clutched mugs of tea to their chests as they watched us, fully loaded and wearing everything we had, crunch our way through the frost to the road.

This was it.

The final push.

In two hours, we would have made it back to London.

For the last nine months we had written, blogged, tweeted and talked of solar power. Equipped with the latest in custom built nano-solar technology we had been waved off by the Mayor of London, received medals in the South of France, crossed deserts, been blinded by concentrating solar power stations, videoed for the 'We Support Solar' campaign, seen the largest PV panel in the world, learnt about renewable energy legislation and presented at a week of events promoting solar power and cycling. Followed by the Evening Standard, reported by the Guardian and profiled in numerous environmental websites and publications, our mission had raised thousands of pounds for SolarAid and would now see us back where we started, largely unscathed from a 13,500 mile pedal-powered circumnavigation of the world.

It had been a total blur.

"Suse?"

"Yeah."

"You know we've been traveling together for a while now."

"Yeah."

"And you know that you don't always listen to me."

"I do! Ha ha—ok.... yeah."

"Well, there's something I really need to tell you."

"There is?"

"Yep—and it would be good if you actually listened."

"Really?"

"Yeah—cos it's important."

"Important?"

"Yeah."

"Ok. What is it?"

"Well, it's just that.... Suse.... you're on the wrong side of the road!"

City Hall, London, England   **21952.63km**

SolarAid

SolarAid confronts humanity's greatest challenges—climate change and poverty—by harnessing humanity's greatest resource, the extraordinary power of the sun.

Established by entrepreneur Jeremy Leggett in 2006, the organisation currently works in Tanzania, Kenya, Zambia and Malawi, delivering effective projects with enterprise at their heart.

There are 1.6 billion people in the world who have no access to electricity, yet many of the poorest communities in the world receive the highest levels of sunshine known to man.

Through its social enterprise, SunnyMoney, SolarAid sells solar lights to poor communities who have an abundance of sun; freeing them from a dependency on kerosene lamps—which are brutally dangerous, polluting and cripplingly expensive.

Kerosene, which remains the primary source of lighting in rural Africa, drains an astonishing 3% of the world's oil supplies and can cost a family living in poverty a large proportion of their income. More alarmingly still, the burning of this filthy oil contributes to indoor air pollution that kills more people across the world than malaria.

Since 2006, SolarAid has sold almost 50,000 solar lights and reached nearly 1 million people with clean, safe solar energy. But the potential to have an even greater impact is enormous.

By helping to create a distribution network, making solar lights available in every corner of the developing world, this ambitious charity is working towards a huge goal: to banish the kerosene lamp from Africa by the end of the decade.

To join them in this mission, or to simply find out more go to:
www.solar-aid.org

DESERTEC

"Within six hours, deserts receive more energy from the sun than humankind consumes within a year."

The DESERTEC Foundation was founded on the 20th January 2009 as a non-profit foundation with the aim of promoting the implementation of the global DESERTEC Concept "Clean Power from Deserts" all over the world.

In 2009, the non-profit DESERTEC Foundation founded the industrial initiative Dii GmbH together with partners from the industrial and finance sectors. Transgreen was founded in July 2010 within the framework of the Mediterranean Solar Plan of the Union. This industrial initiative is aimed at promoting the construction of power transmission lines in the region, complementing the work of the Dii.

The DESERTEC Foundation also launched the DESERTEC University Network as a platform for scientific and academic collaboration.

DESERTEC demonstrates a way to provide climate protection, energy security and development by generating sustainable power from the sites where renewable sources of energy are at their most abundant. These sites can be used thanks to High-Voltage Direct Current transmission.... with losses of just 3% per 1,000km. Thanks to heat storage tanks, concentrating solar-thermal power plants in deserts can supply electricity on demand day and night.

As 90% of the world's population lives within 3,000km of deserts, DESERTEC can be realized not only in Europe, the Middle East and North Africa (EU-MENA), but also in Sub-Saharan Africa, Southern Africa, the Americas, Australia, India and in the whole of East Asia where the centers of demand are within the reach of suitable deserts.

More details can be found at: www.desertec.org

Ten Lessons for the Long Distance Cyclist

1—You will ride through road kill

With cliffs or fences on one side of you and fast moving traffic on the other, veering around dead animals is not always possible. The most gruesome of these experiences will be when you pedal through the putrefying carcass of a decomposing camel.

2—Traffic is never predictable

No matter how rehearsed you are at riding down busy roads at home, this will not prepare you for the moment a cow nonchalantly sidles into your path. Ditto: tuk tuks, buses, lorries, fruit sellers, bullock carts, rickshaws, hawkers, dogs, cats, chickens, monks on motorbikes and a family of seven carrying a pig.

3—You will get lost

Even with a map, GPS and an encyclopaedic memory, you will still get lost.

4—You will get groped

Even in un-dramatic circumstances you will be hugged, kissed and have your boobs fondled by a barrage of elderly matrons [women]. Or have your thighs squeezed, wrestled and given unexpected endings to a muscle relieving massage [men].

5—Weather hurts

You will be blown, scorched, frozen, buffeted, pelted, melted, drenched and burnt.

Fact.

6—You will be kidnapped

a) by those intent on housing you
b) in a vodka-based hostage situation, where your cunning captor will ensure you are too hung-over to leave
c) properly

7—A map is the best bit of kit

You can use it to plot, plan and to navigate. You will flash it at police, point your route out to soldiers and use it as a way to communicate with a plethora of interested bystanders. Everybody loves a map.

8—Your body will be abused

You will not eat enough, not drink enough and/or get food poisoning from an errant clam.

9—You will have irrational thoughts

If you are traveling alone you will, at some point, truly believe that you are an international counter terrorism expert. If you are traveling with others you will, at some point, want to push them off their bike and down a steep ravine. Especially if they tell you ONE MORE TIME HOW EASY THEY ARE FINDING IT.

10—You will, without trying, have the strangest most randomly epic adventures of all time.

So get on your bike.

"Adventure is what you make it; and, whether it's the travel, the discovery, or just the feeling of letting go, the only way to find out is to go out there and do it."—Travis Rice